UA国际 建筑设计 2000-2012
SELECTED AND CURRENT WORKS

UA国际 著

天津大学出版社

图书在版编目（CIP）数据

UA国际建筑设计2000－2012：英汉对照 /
UA国际编著. -- 天津：天津大学出版社，2012.9
ISBN 978-7-5618-4446-5

Ⅰ．①U… Ⅱ．①U… Ⅲ．①建筑设计－作品集－中国
－现代 Ⅳ．①TU206

中国版本图书馆CIP数据核字(2012)第195088号

责任编辑：朱玉红
美术编辑：朱俐倩

出版发行　天津大学出版社
出 版 人　杨欢
地　　址　天津市卫津路92号天津大学内（邮编：300072）
电　　话　发行部：022—27403647　邮购部：022—27402742
网　　址　www.tjup.com
印　　刷　上海锦良印刷厂
经　　销　全国各地新华书店
开　　本　270*270mm
印　　张　33
字　　数　318千
版　　次　2012年9月第1版
印　　次　2012年9月第1次
定　　价　398.00元

致 谢
ACKNOWLEDGEMENT

中国的经济发展和城市化进程令世界瞩目。近20年来，建筑设计行业得到了前所未有的发展机遇，建筑师也得到了很多实践的机会，而UA正是伴随着这一进程诞生并成长的。

感谢这个时代！身处此时此地是我们这代建筑师的幸运。

感谢绿地集团、保利集团等一批有社会责任感的企业！我们有幸在与你们的合作中近距离感到你们的激情和使命感。是你们为中国的城市面貌带来了翻天覆地的变化，为社会创造了巨大的财富，更为我们提供了广阔的创作天地，让我们的想象得以落地实现。

感谢诸位甲方建筑师！国际通行的"建筑服务"需要负责从项目的设计、施工直至交付的全过程，包括质量、进度、成本控制和合同文档管理；其中职业建筑师的职责包括设计和建造活动的管理（督造）两方面内容。但中国现行的设计管理习惯和工程监理体制将建筑师的职责一分为二：由甲方建筑师和乙方建筑师分别承担。甲方建筑师充分了解资本的需求并熟谙建造的管理，乙方建筑师负责设计，这种甲乙方建筑师并置且无间合作的体系符合中国的市场现状，并创造出令众多发达国家无法企及的效率和成就。因此，没有甲方建筑师的智慧和汗水，任何一个设计作品都无法完美呈现。

UA谨以此书向各界朋友致敬。我们一直努力在各种因素的制约下，使自己的每件作品都富有想象力和高品质，在此呈现的仅仅是UA众多作品中的一小部分。对时代常怀感恩之心，对设计永存敬畏之情，这是UA人永远不变的情怀。

Fast economic development and urbanization of China have stunned the world. In recent two decades, architectural design industry has met unprecedented development opportunities and architects have also got many chances for practice. UA was founded and grows up under such background.

We thank this era because we are lucky to be here and witness all these surprises.

We shall give our acknowledgment to Greenland Group, Poly Group and other enterprises taking on their social responsibilities! We are honored to feel your passion and responsibility in our close cooperation. You have brought so many changes to Chinese cities, created huge fortune for the society, provided wide creation space for us and enabled us to realize our imagination.

We also want to show our acknowledgment to all owners' architects! "Architectural service" which is generally accepted in the world shall be responsible for all processes from project design and construction until completion and delivery, including control of quality, schedule and cost, as well as management of contract document; professional architects are responsible for design and construction management (supervision). However, current Chinese design management habit and engineering supervision system divide architects' responsibilities into two parts, which are taken by owner's and construction unit's architects respectively. The owner's architect has sufficient understanding about capital demand and mastered construction management, while construction unit's architect is responsible for design. Such close cooperation between owner's and construction unit's architects meets the status quo of Chinese market, and has obtained outstanding effects and achievements which have not been realized in many developed countries. Therefore, no perfect design work can be created without the wisdom and endeavor provided by owner's architects.

UA shows, in this book, its sincere respect to partners of all fields. We have done and will continue to do our best to infuse imagination and high quality into each of our design work and only a part of our works are selected and collected in this book. UA staff's tenet is to show constant respect to the era and design task.

目 录
CONTENTS

前 言 PREFACE   006

1 超高层建筑 SKYSCRAPER   011

2 酒店及城市综合体 HOTEL AND URBAN COMPLEX   037

3 文化教育建筑 CULTURAL AND EDUCATIONAL BUILDING   095

4 居住建筑 RESIDENTIAL BUILDING   173

5 城市规划及城市设计 URBAN PLANNING AND DESIGN   347

前言
PREFACE

新世纪的第一个10年，我们经历了世界经济的不稳定性、国际政治的波动、新技术的发展以及能源危机，建筑师目前站在新世纪的十字路口，面临更多从未预期的问题，如何正确地提出问题远远比对现有问题的解答更为重要，如何实践建筑成为一个重要的话题。法国当代著名思想家米歇尔•德•塞都(Michel de Certeau)在他的《日常生活实践》(The Practice of Everyday Life)一书中对于战略与战术有精彩的引用和描述。德•塞都着重关注了在社会学意义中，人们每天是如何在已经规划完善的城市及其生产物中寻求变通的，比如在井然有序的街道中进行各种活动，或者在法律体系中运用语言使自己获得利益。也就是说，在由社会科学编织的传统、语言、符号或者艺术不断产生交流的时候，人们有充足的机会去扰乱这个系统，重新寻找使用系统的可能性，这便是文化产生的缘由。在通常认为的"实践"过程中，这点往往被忽略。

德•塞都指出了"实践"过程中的两种不同的行为，即战略与战术。这两个引用自军事的术语在此书中被成功地赋予了全新的社会学意义，并具有可操作性。美国普林斯顿大学建筑学院院长斯坦•艾伦赋予战略和战术以建筑学的意义：战略是提前进行的规划的行为，它与现场发生的事件保持一定的距离；而战术则是在现场进行的即时的策略选择，它表现为对不断改变的现场条件所做出的反应。战略所涉及的工作具有一定的权威力量，目的在于掌握、控制并完成目标，具体则是关于如何计算结果，预知未来，建立系统结构等的工作。完成这项工作必须具有相当丰富的知识，仅仅靠建筑师的参与，则相对力量单薄。因此，建筑师自身必须尽可能地具备更多的相关知识，并且学会如何与不同知识领域的专家合作。战术则更多针对具体问题，它具有"即兴的、不确定的、由直觉及经验所决定的"等特征，因此它不得不涉及时间尺度，在现场进行机动的调度与调配。对于建筑师来说，在解决客户需求及建中面临各种问题。建筑师固然会面对现场的诸多不确定条件，但建立有效准确的战略却可能是建筑师需要并且正在做的工作。

UA国际成立于世纪之初，经过了10余年的磨砺，由一个小型建筑设计事务所的规模，经历了若干次飞跃，目前已经成为一个拥有8个方案设计所、1个综合甲级设计院、约500位建筑师，综合产值和人均产值均在全国名列前茅的民营大型综合设计机构。可以说UA从战略上和战术上都是非常成功的，UA也因而成为强有力的国际设计品牌。

During the 1st decade of this century, having experienced uncertainty of global economy, disruptive world politics, development of new technology and serious energy crisis, architects are standing at a crossroad. We are facing more unexpected problems, thus how to raise questions is more important than to answer current ones. How to practice architecture becomes crucial. In his book *The Practice of Everyday Life*, the contemporary French philosopher Michel de Certeau had amazing discussion on strategies and tactics. De Certeau states, in Sociology, how people in daily life look for change in already planned cities and artifacts, e.g., to benefit through various activities in ordered streets or through language in law system. In other words, when tradition, language, symbols or arts, which are ordered by social sciences, constantly interact, people have enough opportunities to interrupt the system in search of new possibilities using the system. This is how culture was created. However, it is usually neglected in the process of "practice".

De Certeau distinguishes two types of "practice": strategies and tactics. In his book, these two military terms have new sociological meanings and are feasible. Stan Allen, Dean of School of Architecture at Princeton University, gives strategies and tactics architectural meanings. Strategy is a carefully devised plan of action to achieve a goal, and it keeps a certain distance from current event; Tactic is a selection of strategy in order to achieve an immediate aim and is a response to on-site ever-changing conditions. Strategy involves subjects having certain authority power and its purposes are to master, control and achieve the devised aims. To be specific, it is about how to calculate result, predict future, set up system, etc. This job requires rich knowledge. Only with architects' participation it is not powerful. Thus, architects must learn more related knowledge as possible as they can, and learn how to cooperate with experts from different fields. On the other hand, tactics are about more specific questions. It is spontaneous, uncertain, and decided by intuition and experience. So it has to do with temporal dimension and distribution and arrangement on site. For architects, they face problems while solving client's needs and during construction. Although architects face many uncertain conditions, they need to establish efficient and precise tactics.

Founded at the beginning of this century, experiencing several leaps and bounds through the first decade, Urban Architecture (hereinafter referred to as UA) has transformed from a small-scale office to a large private architectural design firm with eight scheme design institutes, one general grade-A design institute and about 500 architects. UA is ranked in the first tier in total output and per capita output among private firms in China. UA has achieved a great success both strategically and tactically and it has also become a strongly competitive international design brand.

城市·建筑

正如UA的名字所示，Urban（城市）+ Architecture（建筑），这两个关键词贯穿UA整个设计及管理的各个部分。U与A是两个对建筑师最基本的概念，两种不同的尺度和思维不可分割，以城市的视野控制建筑，以建筑的工具塑造城市。UA的作品遍布于全国24个省级行政区的50个城市，包括超高层公共建筑逾15幢，大型城市综合体逾20座，年设计建筑面积超过1000万平方米。在城区，他们重新塑造城市中心的形象、空间、生活；在郊区，他们建起新的城区，新的生活模式。大型地产商如绿地、保利、复地、万科、中海、世茂、远洋、长甲等均与UA建立长期的合作关系，在为业主提供良好服务的同时，UA的设计作品也取得了非凡的市场成功。他们的设计作品在2009、2010、2011年连续3年成为上海单盘销售冠军，2010年为北京、成都单盘销售冠军。UA创造出市场、服务、技术、艺术、理念高度综合的和谐共存。

过程·产品

UA之所以成功，在于其创立之初，便关注产品的各个环节。设计之始需要大量对产品的定位、类型、模式、概念、地域性的多层次调研与深度分析，以达到最佳最有效的成果。每一步的战略与战术都寻求完美，这体现在UA特有的"四会定案"制度。专业委员会是一个个特殊的部门，他们管控每一个项目每一步的品质，四个不同委员会（规划、产品、造型、施控）由公司内各领域的专家和设计高手组成，对每一个项目从以下四个层面进行单独上会评审：1.规划的合理性与前瞻性；2.产品单体的功能布局与细节；3.造型设计的艺术性、创新性、经济性和可操作性；4.施工控制，包括建筑单体细节的二次设计。通过四会审核才能发至业主。对通过四会的优秀项目进行表扬和公开展示，需要修改的要重新上会再次审核，或者重做。这使得UA不单单是一个公司，更是一个高效的研究机构与"大学"。高效、务实、严谨、品质是UA的核心。UA人经常说道："UA对作品本身的珍视程度犹胜于我们的业主，所以每一个项目我们必将倾注最大的心血去打造。"UA庞大的材料样品库为施工控制提供强大基础，但富有经验的建筑师还是花费大量时间进行现场控制，每每总是让人意外，建成的效果常常超过效果图的精美。对过程和成品的高质量把控成就了UA的高度。

URBAN / ARCHITECTURE

Just as the UA's name shows, Urban + Architecture, these two key words run through the whole design and management processes in each part of UA. For architects, U and A are the two most basic concepts, which are two indivisible different scales and ways of thinking, to create architecture in urban scale and to form the city by architectural means. UA's projects spread over 50 cities in 24 provinces, including over 15 super-high public buildings and more than 20 large urban complexes with the annual design floor area exceeding 10,000,000 square meters. In urban area, they renew the image, space and life in urban center. In suburban area, they create new town area and lifestyle. After establishing a long-term cooperative relationship with high-end real estate developers such as Greenland Group, Poly Group, Forte, Vanke, Zhonghai (Chinese Overseas Property), Shimao Property, Sino-Ocean Land and CJ Land, UA's works become the best-selling individual project in Shanghai for three consecutive years from 2009 to 2011 and in Beijing and Chengdu 2010. UA has now successfully created a high-comprehensive harmony between architectural market, service, technology, art and concept.

PROCESS / PRODUCT

The reason why UA has achieved such great success is that they pay close attention to every step of design procedure, from the beginning of its founding. In order to achieve best effect, they work on intensive multilevel researches and profound analysis about identification, typology, modes, concept and regionality. This pursuit of perfect strategies and tactics is well represented in the UA's unique system of "Four Committees Determine Schemes" (which means that every scheme is jointly determined by four different committees). Discipline committees are special departments and they strictly control each step of every project. Four different committees (for planning, product, appearance, construction control) are composed of experts and design masters from different fields in UA. They are in charge of each following part of every project: 1. reasonability of planning and farsightedness; 2. functional layout and details of single product; 3. artistry, innovation, economy and feasibility of appearance design; 4. construction control, including secondary design of detailing of each individual building, which is examined by four committees respectively. Only the proposals which are verified and qualified by four technical committees can be sent to clients. These projects with high scores will be praised and exhibited to the public while the ones with low evaluation are required to be modified or even redesigned. UA is not only a design firm, but also an efficient research institute and "university", which has the core ideas focusing on high-efficiency, pragmatics, rigorous attitude and high-quality. UA's staff always say: "We treasure our own products more than our clients do, so we devote tremendous energy to every project." UA's large material sample library is a strong base of construction control. However, experienced architects also spend considerable time in field control. It is amazing that when building construction is completed, the real buildings are always more appealing than their previous computer renderings. This high-quality control of process and product makes UA reach a new height.

设计·创新

优秀的建筑设计特别是房地产类的设计绝不是一时的灵感闪现就能成的，它基于大量繁杂而细致的收集整理、研发总结和提炼提高。这些研发成果以及在研发基础上的创新成果，是UA诸多设计实践得以成功的根基。特别是在中国当前阶段，经济快速发展并伴随着高速的城市化，在研发成果基础上的设计是保持建设速度、保证建设质量的最有效手段。这个研发创新的过程在UA一直持续存在着，UA也一直骄傲于自己是一所学院型企业，并通过研发创新成果为业主创造更高的价值。

UA目前集中在超高层建筑、城市设计及居住区规划、商业及城市综合体、创新居住产品等范畴进行研发，并已形成200多项研发成果。其中如"三代开放社区规划模式研究"、"低层豪宅底层地面系统研究"、"公寓小户型及公共部位精细化设计指引"等独创理念，在实际应用中都体现出巨大的效能优势。

UA所关注的城市与建筑，并非只是物质的和形式的，它更是生活的与感知的。设计不仅仅只是风格与样式，他们在设计之初首先提出生活的模式和使用的效果。维特鲁威提出的建筑三要素"实用、坚固、美观"中，"实用"是首位的，这样设计作品才能深入人心，被开发商和使用者欣然地接受与赞美。风格不是简简单单的复制与山寨，现场看UA的设计，很难说它们是严格地照搬法式、西班牙式、美式，而是基于使用需求、市场走向、地域文化等综合的成果，他们每每发展出新的"风格"，立即成为其他公司竞相模仿的对象，但在被模仿之后他们又发展出了新的"风格"，走在别人之前需要多么强大的思考、能力和技术啊！所以，UA的住宅和商业项目总是让人竖起大拇指，获得开发商和业主的一致好评，物有所值。

在住宅和商业设计的高峰之后，他们没有将公司简单地定义为商业公司，而是一个多类型多尺度的综合性机构。经过多年的积蓄，在多个超高层项目中标后，他们建立起强大的极专业的超高层设计团队，公司走向一个更高的平台，拥有更好的前景。

DESIGN / INNOVATION

The excellence of architectural design, especially that of real estate projects, is absolutely not obtained from a flash of inspiration, but based on a great number of multifarious work on collection, arrangement, research, summary and elevation. Based on these researches and relevant innovation achievement, UA has gained successful design practice. Currently China is in the rapid development of economy and accelerated urbanization, thus the design from research is the best tool to keep the speed and quality of construction. This kind of consistent process of research and innovation always exists in UA and UA has been always proud of being an academic firm and creating higher value for clients through research and innovation achievements.

Currently, UA focuses on skyscraper, urban design, residential planning, commercial complex and HOPSCA, innovational residential product. There are more than 200 research projects for efficient superiority in practice. UA's many creative concepts, such as "Research on 3G Open Community Planning Form", "Research on Floor System of Ground Floor in Low-Rise Luxury Residential Building", "Guide for Detailed Design of Small Unit and Public Area in Apartment" have showed huge efficient advantages in practical application.

Urbanism and architecture that UA concerns are more than physical and formal but also lifestyle and sensibility. Besides just architectural style and form, their design pursues new lifestyle and offect of using at the very beginning of design. Three fundamental principles of architecture proposed by Vitruvius can be summarized as utilitas, firmitas, venustas, among which the first one is also the most important, thus, the projects can be accepted and deeply rooted among the clients and users. Style is not just about copy and simulation, but based on a comprehensive outcome from needs, market and local culture. It is hard to say that UA's projects are just strict copies from French, Spanish and American styles. Every new style created by UA often causes a wave of imitation, then they develop new style afterwards. What great idea, ability and technology are required for them to hold the lead! So, almost all the residential and commercial projects of UA are highly praised by both users and clients.

Reaching the peak of residential and commercial design, they did not simply define UA as a commercial firm, but a multi-type and multi-scale comprehensive institute. After winning the bidding of several skyscraper projects with the many years' accumulation, they have built up a considerably professional team upon skyscraper design, which promotes the firm to a larger platform and great prospect.

文化•管理

UA经过多年的积累，形成了独特的设计与管理文化。三位创始合伙人的人格魅力和综合能力都令人钦佩，他们之间又十分互补，使得公司更上一层楼。他们创造出在小型事务所和大型设计院之间的一种新模式，8个不同的方案设计所和1个施工图团队紧密结合但又相对独立运作。好的公司不只是一个商业的机构，它以人为本，对员工进行充分的技术实力的培训、人文修养的补充、团队凝聚力的打造，这些都是对UA成功强有力的支持。

战略•前景

UA的决策者总是走在时间的前面。2008年经济危机之时，其他公司纷纷裁员，他们却不减反增，储备了大量的优秀设计师，待市场回暖之时，他们又必然地经历了质的飞跃。2011年，在10周年之际，经过半年多的改造与精心设计，UA的各个所搬进了独具特色、独栋且连贯的黑楼、红楼、绿楼、黄楼、玻璃楼等六栋独立办公楼，各个团队之间的竞争、特色、合作更为突出。除了建筑方案设计和施工图，UA目前也建立起强大的景观室内设计团队，向更多元的方向迈进。UA的第一个10年与新世纪的第一个10年共同发展，他们提出问题的方式，让解答更为轻而易举，他们高瞻远瞩的战略与灵活应变的战术都让我们更为期待后几个10年的UA！

王飞
香港大学建筑学院
2012年5月

CULTURE / MANAGEMENT

After many years of practice, UA has formed its unique culture of design and management. The three founding partners all have praiseworthy personal charms and comprehensive abilities. Their complementary characters helped UA reach a higher level. UA runs a new model between small design studio and large design institute, and the eight scheme design departments and one construction drawing team have tight bonds but run independently at the same time. An excellent corporation is not only a commercial organization but also an aggregation of individuals, which puts its staff in first place. It focuses on training the skills of staff, improving the cultural level of people, and building up the team cohesion, which have made UA so successful.

STRATEGY / PROSPECT

The decision makers of UA always walk ahead of time. During the 2008 Economic Crisis, they hired many of high-level designers while many other companies reduced the staff . So, when the market gets warm again after a cold spell, UA gains a foregone leap. After more than half a year's design and refurbishment, all design departments moved into the detached and also connected six office buildings such as black-house, red-house, green-house, yellow-house and glass-house at its tenth anniversary in 2011. The characters of different teams and the competition and cooperation with each other are even more standing out. Besides architectural scheme design and construction drawing, UA has built up powerful landscape interior design teams, pursuing more diversified development. The first decade of UA developed with the development of the new century, the way they raise a question makes the answer easier, and their foresighted strategy and flexible tactics make us anticipate more for the coming decades of UA.

Wang Fei
Faculty of Architecture, The University of Hong Kong
May 2012

1 超高层建筑
Skyscraper

超高层建筑是人类征服和改造自然的图腾式表现。近年来，随着中国经济的快速发展和人们越来越多地掌握超高层建筑的建造技术，超高层成为中国城市化进程中最为耀眼的一种建筑表现形式。超高层设计内涵丰富而广泛，需要设计团队对地域文脉、城市关注、建筑美学、结构体系、机电系统、幕墙技术、垂直运输系统、建筑物理、生态技术和计算机辅助设计等具有高度的整合和把控能力。UA近年来持续关注中国超高层建筑的发展，并积极参与到诸多设计实践中，逐渐形成了强大而专业的设计团队和知识优势，现已成为中国少数具备大型超高层建筑综合设计能力的设计机构之一。

Skyscraper is a totem-type representation for human to conquer and transform the nature. In recent years, with fast development of Chinese economy and emergence of more and more skyscraper construction techniques, skyscraper has become an eye-catching architectural manifestation during urbanization in China. Skyscraper design has comprehensive and extensive connotation and requires design teams to possess high-level integration and control abilities in the aspects of local culture, urban concern, architectural esthetics, structural system, electromechanical system, curtain wall technology, vertical transport system, building physics, ecological technique and CAD, etc. In recent years, UA has paid constant attention to the development of skyscraper in China and has actively participated in many design practices, gradually forming its own powerful and professional design teams and know-how. At present, UA has been one of few design organizations possessing comprehensive design ability of large skyscraper in China.

BEIJING GREENLAND CENTER PLOT 625

北京绿地中心（625地块）

用地面积	建筑面积	设计时间	建筑类型
33.21万平方米	60.25万平方米	2010年9月至今	商办+高层住宅

基地南边是望京中环路，北至大望京二号路，紧邻大望京商务区的带状公园，东至大望京三号路，西临新望京干道。地块的控制高度300米。项目用地附近有轨道站点，位于用地的西北，公园内部。

设计采用相对独立的块状造型，力求将建筑独特的功能模块互相连接整合为一体，将一座典型的北京"超级大厦"分解成互相联系的几个部分，在疏解基地内人行交通的同时与周边城市环境紧密衔接，加强了同既有建筑的联系。北方地域的气候特点与大气磅礴的审美情趣造就了北京建筑体量庞大而又简洁的风格；同时皇家建筑中精美细腻的木作技艺又赋予了这些建筑动人的细部。受此启发，设计不再沉湎于寻找表面的美丽和矫揉作态的丰富感，而是寻求洁净的、直截了当的美，这样的美更加真实。事实上，我们理解的"简约"并不是单纯控制成本的直接结果。相反，它恰恰是丰富的集中统一，是复杂性的升华。这还意味着无须更多不必要的立面装饰元素，取而代之的是经过谨慎选择的形式和材料，极其坚固和安宁的完美结合。

Beijing Greenland Center (Plot 625) is orbit the middle-ring road of Wangjing Subdistrict orbit, No. 2 road of Dawangjing at north, No.3 road of Dawangjing orbit and new trunk road of Wangjing Subdistrict at west. In addition, it is also near to a strip-type park in Dawangjing business district. Controlled elevation of the plot is 300m. Surrounding orbit stations are located at the northwest of the plot and in the park.

The design adopts relatively independent block shape to integrate unique functional modules of building together and divides a typical Beijing-style "super skyscraper" into several parts which are mutually connected. Thus, traffic load on the plot is alleviated, so as to link the plot with surrounding urban environment more closely and to strengthen the connection with existing buildings. Climatic characteristics of Northern China and generous aesthetic concept create large and simple style of buildings in Beijing; meanwhile, delicate timber technology of royal building infuse charming details to these buildings. Inspired by these aspects, designers don't indulge in realizing beautiful appearance and complicated structure any longer, but pursue clean and simple beauty which gives people truer feeling. In fact, the "simplicity" we understand is not the direct result of simple cost control. In contrast, it is no other than comprehensive integration and a kind of upgraded complexity. Moreover, it also means that unnecessary decorative elements on facades are replaced by selected forms and materials, which realize firm and peaceful aesthetic integration.

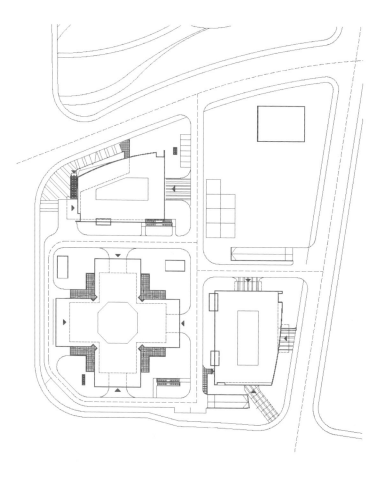

above site plan/总平面图
top right A,B,C tower floor plan/A,B,C座楼层标准平面图
bottom right bird's eye view/鸟瞰图

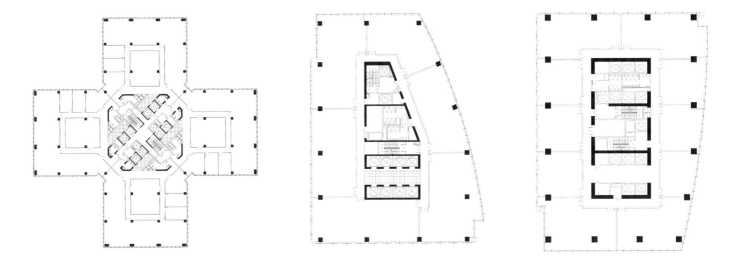

left exterior view/外景图
above entrance exterior view/入口外景图
below section diagram/节点图解

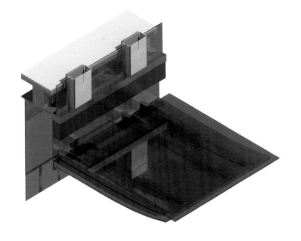

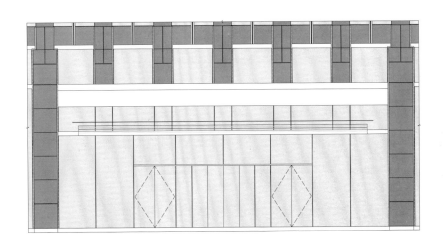

left and below exterior view/外景图
top right section diagram/节点图解
bottom right CAD diagram/CAD图解

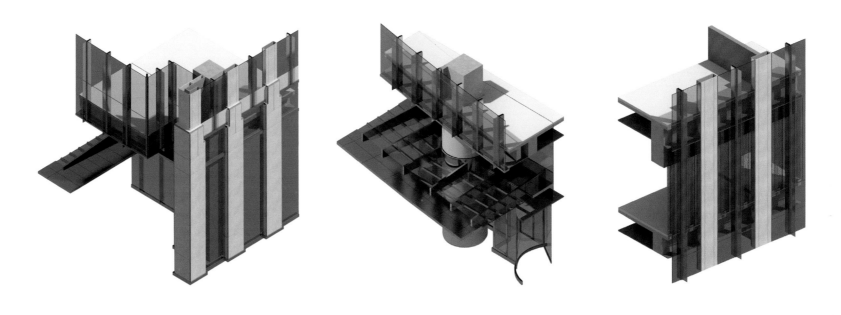

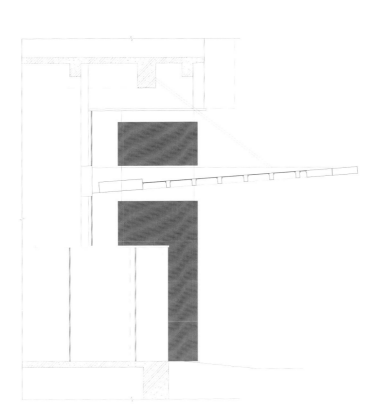
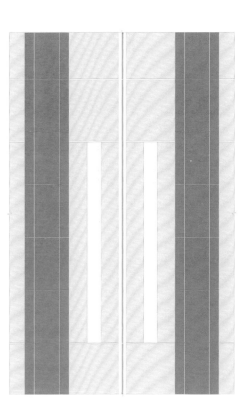

XI'AN GREENLAND CENTER

西安绿地中心

用地面积	建筑面积	设计时间	建筑类型
17 000平方米	165 000平方米	2011年1月至今	超高层

十三朝古都西安正经历一次新的发展大浪潮，以绿博会为契机大力发展开发区经济。西安绿地中心正是这大浪淘沙中一颗闪亮的钻石。东西两座双子塔顶高270米，是开发区乃至整个西安市的天际制高点。从远处看犹如一座巨门，面向开发区CBD的建筑群。整体造型以钻石为原型，强调精准切割的体形美和玻璃材质所体现的璀璨质感。摒弃多余的装饰，只采用简洁的形体切割手法，意在充分强调接近300米高度的体形，和形体本身所表现出的完美的比例。

Xi'an City is an ancient capital of 13 dynasties and is witnessing a new development surge. It grasps the opportunity of Green Expo to realize fast economic development in the development zone. Xi'an Greenland Center is just a shining diamond in this development surge. The two twin towers located at east and west are 270m high and they form the highest skyline in the development zone, even in Xi'an. Viewed from a far distance, they are like a huge gate opening up to the building cluster in CBD of the development zone. These twin towers are designed based on the shape of diamond, emphasizing on the precisedly-cut aesthetic appearance and glistening glass material. The method of replacing redundant decoration with simple shape highlights the unique building shape about 300m high and perfect dimensional proportion of building body.

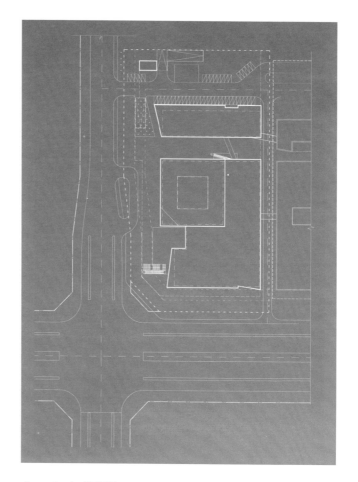

above site plan/总平面图
right rendering view along Jinye Road /沿锦业路效果图

below　facade rendering view/正立面效果图
right　tower plans/塔楼平面图

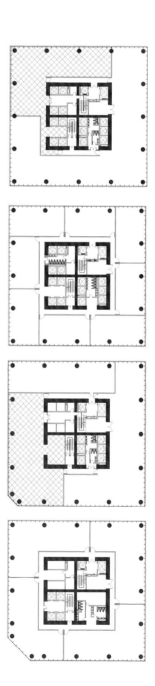

above   concepts contrast/多方案对比
below   form polish/形体推敲

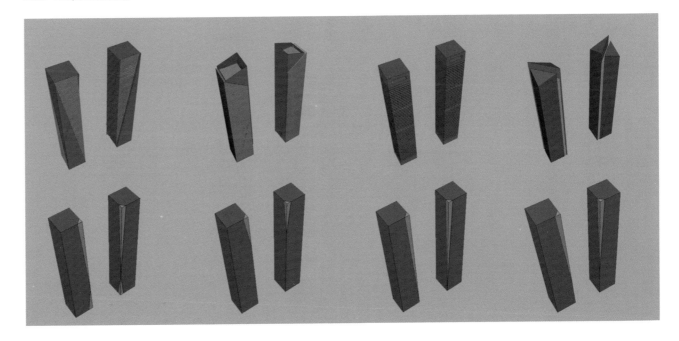

DAQING HIGH-TECH ZONE FINANCIAL PARK

大庆高新区金融产业园项目

用地面积	建筑面积	设计时间	建筑类型
25.52万平方米	62.59万平方米	2011年11月至今	超高层

　　大庆高新区金融产业园项目位于大庆市龙凤区心脏地带，以办公、五星级酒店、银行、商务楼为主，辅以高档公馆社区。规划方案以一条中轴线主导了地块全局，同时结合外向性的景色优美的市政公园，奠定超高层塔楼的核心位置，并且通过设置内向性的中心公园，将高档公馆社区与商办区域自然分开，同时又共享其独特景观。整体规划在城市中展现出现代尖端独特的金融商务区的气质风貌。

　　超高层双塔高度为216米，建成后将成为黑龙江省最高的双塔建筑。建筑外墙表皮形似鳞片的效果，通过顶部肌理与尺度的微妙变化来实现整体比例的均衡，创造出简洁但优雅的城市地标。

Daqing High-Tech Zone Financial Park, located at the heartland of Longfeng District of Daqing City, is mainly composed of offices, five-star hotel, bank and business buildings and is supported by high-end residential community. According to the planning scheme, a piece of axial line passing through the whole site and municipal park providing extroverted and beautiful landscape emphasized the core position of the skyscraper tower. Introvert central park is designed to naturally separate the high-end residential community from business area, while providing unique landscape to both areas. The general planning shows unique and sophisticated characteristics of modern financial business zone.

The super-high twin towers are 216m high and will be the highest buildings in Heilongjiang Province after completion. Buildings adopt scalelike facades and realize balanced proportion through subtle variation of top texture and scale, so as to create a simple but elegant urban landmark.

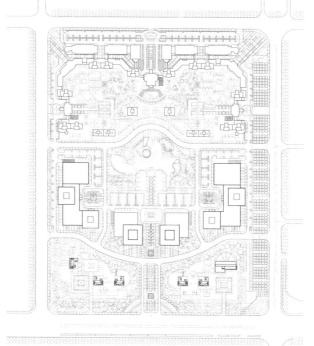

below plan/总平面图
right rendering view along Century Avenue /沿世纪大道效果图

above　concepts contrast/多方案对比
below　tower plans/塔楼平面图
right　conception result/概念成果

left and right exterior view/外景图

DALIAN GREENLAND CENTER

大连绿地中心（南区）

用地面积	建筑面积	设计时间	建筑类型
37 772平方米	272 527平方米	2010年6月	超高层住宅及商住

　　大连绿地中心（南区）项目位于大连市钻石港湾、东港新区CBD核心区。南区地块拟建东北第一高楼，总高518米的绿地中心。地块占地约3.78万平方米，总建筑面积约27.2万平方米。

　　项目整体布局以518米超高层塔楼为核心，塔楼的三角形结构为母题，整体空间形态呈众星捧月之势。建筑整体造型力求简洁，强调建筑几何体量，造型时尚、典雅、大气，在强调形态同时，更强调功能和造型的结合及建筑群组的韵律感。通过对关键节点比例的精确控制使整个建筑现代中透露着古典建筑的韵味。

　　在大连这个浪漫而又严谨的城市，有美丽的海港、波光闪耀的大海、清新的海风、翩翩起舞的海鸥。这些都成了我们设计的元素，建筑处在大连最美丽的海边，如何利用海景海风带来的优势，成了我们立面设计的重点。

Dalian Greenland Center (South Zone) is located at the core area of diamond harbor and CBD of east harbor new zone in Dalian City. Greenland Center with a total height of 518m is planned on the north zone and it will be the highest tower in Northeastern China. The plot covers a total area of about 37,800m² and a gross floor area of about 272,000m².

General layout of the project sets a super-high triangular tower of 518m high at the center to form a spatial shape like a moon surrounded and supported by multiple stars. Efforts are made to realize simple building shape, emphasize geometric volume and create modern and elegant appearance. Integration of reasonable function and structure and rhythmical image of building cluster is carefully concerned while realizing beautiful appearance. Accurate control of proportion among key nodes integrates both modern and classical architectural characteristics into the whole building.

Dalian is a romantic and well-planned city, having beautiful harbor, shining sea, fresh sea breeze and flying sea gulls. These objects have become our design elements. The project is located at the most beautiful seaside of Dalian, so how to utilize the beautiful landscape is a key point in our facade design.

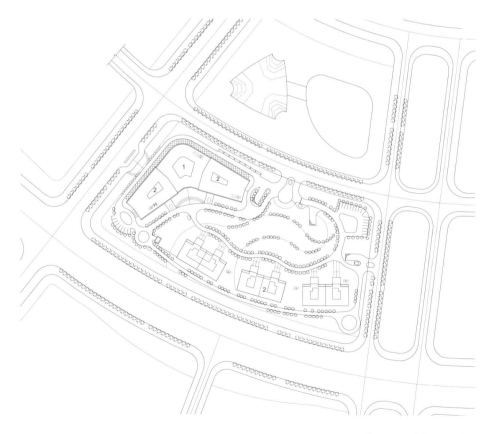

below　site plan/总平面
right　exterior view/外景

1. commercial building/商业
2. apartment building/公寓

below　bird's eye view/鸟瞰图
middle　details/细部节点详图
right　exterior view/外景

1. architecture form analysis/建筑形体分析
2. a triangle component/三角形构件
3. wavy feature elevation/波浪流线型立面

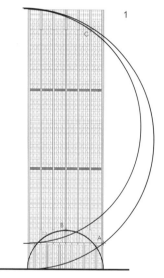

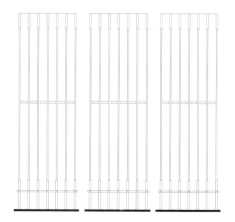

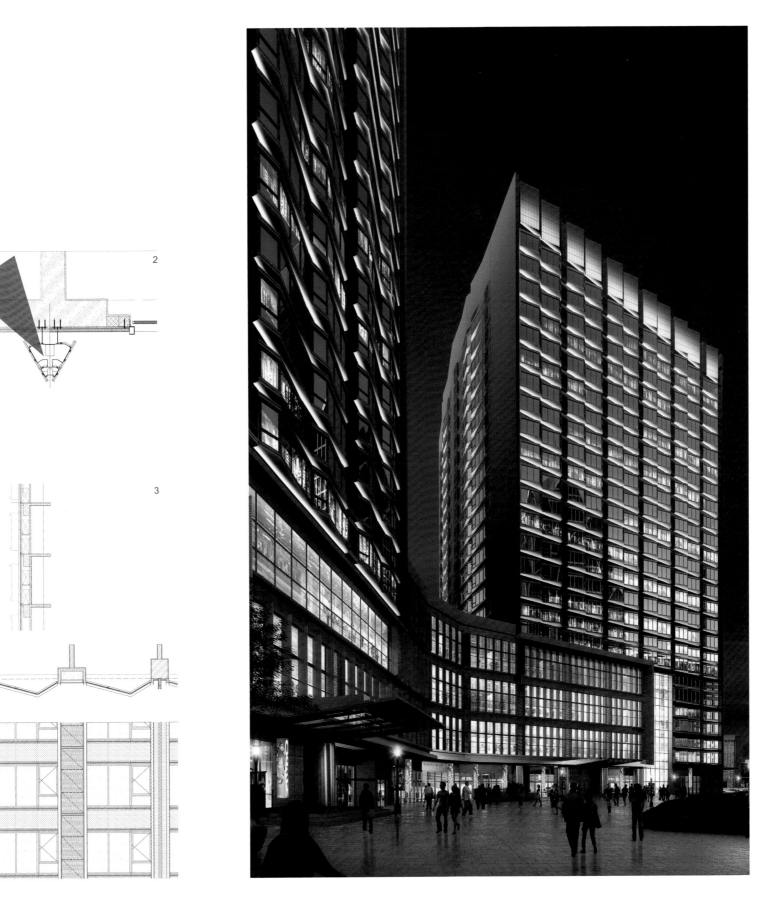

ANQING YINGJIANG CENTURY CITY CENTRAL PLAZA

安庆迎江世纪城中央广场

用地面积	建筑面积	设计时间	建筑类型
48 280平方米	222 238平方米	2010年5月	商务办公综合体

below site plan/总平面
right bird's eye view/鸟瞰图

　　本案位于安庆市新城区内的绿地迎江世纪城20号地块B-1区南侧，向西俯视整个安庆老城中心，西临新区主干道，北起安庆火车站，南至长江，是联系北侧火车站的主要纽带。

　　158米的高度将成为本区域地标性制高点，在整体形态上，本案与地块北侧的生活广场共同塑造了一个和谐完整的商务办公综合体的形象。裙房的建筑设计与场地结合增加了商业展开面，丰富了商业空间，提升了商业价值。交通设计上本着以人为本的原则，有效地组织各种流线，避免车流、货流、人流的交叉。道路结合场地和建筑布局，兼顾场地与周边道路的高差，形成多层次、疏密相间的立体绿化布局是本案场地设计的主要特色和手法。

Yingjiang Century City Central Plaza is located to the south of B-1 zone of plot 20 in Greenland Century City in New District of Anqing City and overlooks the whole old urban center of Anqing in the west. To the west, it is adjacent to the trunk road of the New District, and begins in the north form Anqing Railway Station, ends at the Yangtz River in the south, being a traffic hub linking the railway station in the north.

Buildings of this project are 158m high and will form the highest landmark skyline in this district. The general layout of this project cooperates with the life square at north to create a harmonious and complete business complex. Podium buildings intergrated with the site enlarge the commercial space and increase the commercial value of this project. Traffic design is based on human-oriented principle to effectively organize various traffic systems and to avoid interruption among vehicle, cargo and pedestrian. Road design cooperates with site and building layout and considers the height difference between site and surrounding roads to form diversified vertical green layout with alternate sparse and dense vegetation. This is the main feature and design method of this project.

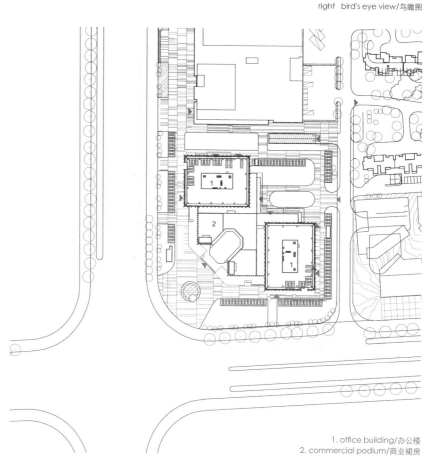

1. office building/办公楼
2. commercial podium/商业裙房

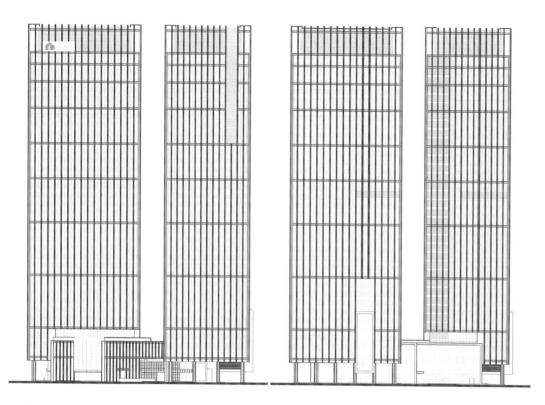

above   south and north elevations/南立面、北立面
below and right   exterior view/外景

2 酒店及城市综合体
Hotel and Urban Complex

城市生活的吸引力在于它的多元与丰富、便捷与高效。城市综合体摒弃了传统单一土地使用性质的开发模式，采用多功能混合开发+立体延伸的开发策略，在有限的土地上整合商业、办公、酒店、娱乐和居住等多种城市功能，具有较强的自我成长、自我繁荣、持续发展的生命力，是实现城市土地高度集约的途径之一，是都市24小时跳动的心脏。UA认为，城市综合体开发的本质是满足服务对象的最终使用目的，城市综合体的最大服务对象是城市及城市居民。因此，城市综合体的开发与设计不仅要突破建筑自身体系的范畴，更要注重其具有的城市意识和社会意识，更要注重使用者的活动和建筑对城市公共生活产生的激励机制。UA擅长于从以下几个方面深入研究城市综合体的设计与运营策略：一、土地能级的提升；二、竖向空间资源的利用；三、多种业态的复合；四、景观环境的优化；五、公共空间的再生。

The charm of urban life is shown in its diversified, abundant, convenient and efficient characteristics. Urban complex abandons traditional development mode of simple land use, and adopts multifunctional development strategy of integrated development plus vertical extension, realizing multiple urban functions on limited land, such as commerce, office, hotel, entertainment and residence, etc. Urban complex has relatively strong abilities for self growth, self prosperity and sustainable development, and it is one way to realize the highly intensive use of urban land and also a 24-hour pulsatile heart of urban life. UA believes that the development concept of urban complex is to satisfy the ultimate use purpose of its users and that the biggest user of urban complex is the city itself and urban citizens. Therefore, development and design of urban complex not only needs to break through the limitation of architectural system, but also shall pay high attention to its urban and social consciousness, as well as the motivation effect generated by users' activity and building on urban public life. UA is good at in-depth research of design and operation strategies of urban complex in the following aspects: 1. increase of land function; 2. utilization of vertical space resources; 3. integration of multiple industrial forms; 4. optimization of landscape environment; 5. regeneration of public space.

BEIJING GREELAND CENTRAL PLAZA

北京绿地中央广场

用地面积	建筑面积	设计时间	建筑类型
40 086平方米	224 202平方米	2009年	商业办公

北京绿地中央广场及绿地缤纷城位于北京市大兴区，紧邻五环，距离市中心15千米左右，交通便捷；地下通道与北京地铁4号线延伸段相连。融合了百货、精品店、大型卖场、影院、娱乐、餐饮等各种时尚休闲的功能，满足了大兴区经济发展和市民购物休闲活动的需求。

本项目的建筑设计充分考虑了人在室外公共活动空间的直观视觉感受和空间尺度体验，营造了宜人的商业体验环境。内凹的广场。下沉的广场等为城市提供了难得的多层次的可停留空间，成为鼓励各种户外活动的场所。建筑群体采用古典式与现代风格并置的手法，形成强烈的视觉冲击感，给人过目难忘的场所感受。

Beijing Greenland Central Plaza and Beijing Greenland Funny City, located in Daxing District of Beijing City and enjoying convenient traffic, are adjacent to the 5th ring road and about 15km from the urban center; underground passage is connected with the extended segment of Beijing Subway Line 4. The project is composed of various fashionable functions, such as shopping mall, boutique, large-scale market, cinema, entertainment and restaurant, etc., so as to satisfy the economic development of the district and citizen's demand for shopping and entertainment.

Architectural design of this project gives full consideration to direct visual feeling and spatial experience in public space, successfully creating appropriate environment for commercial experiences. Concave square and sunk plazas provide rare diversified urban public spaces and encourage various outdoor activities. Buildings integrate classic and modern styles to form an intensive visual impact and to give people memorable spatial feeling.

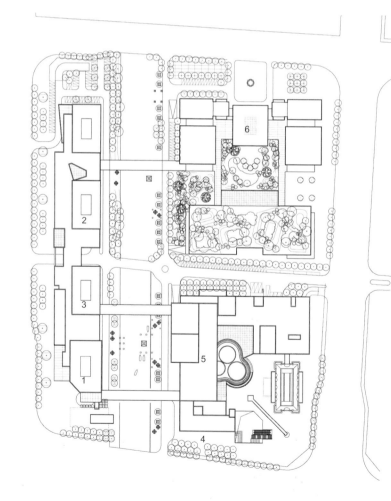

1. quasi grade-A office/准甲办公
2. LOFT office/LOFT办公
3. SOHO office/SOHO办公
4. apartment-style office/公寓式办公
5. shopping mall/大型商业
6. hotels and business facilities/酒店及商务设施

left　bird's eye view/鸟瞰图
above　site plan/总平面图

below photo/实景照
right standard floor plans/标准楼层平面图

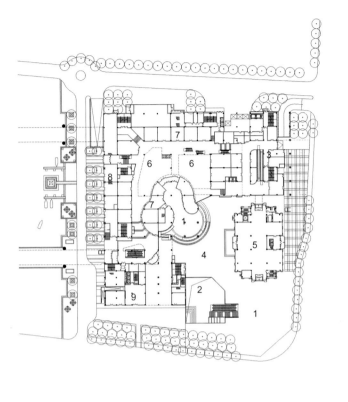

1. commercial plaza/商业广场
2. sunk plaza/下沉广场
3. supermarket/大卖场
4. main entrance/商业主入口
5. clubhouse/俱乐部会所
6. atrium/中庭
7. shop/店铺
8. bar & cafe/酒吧茶座
9. lobby/公寓办公大堂

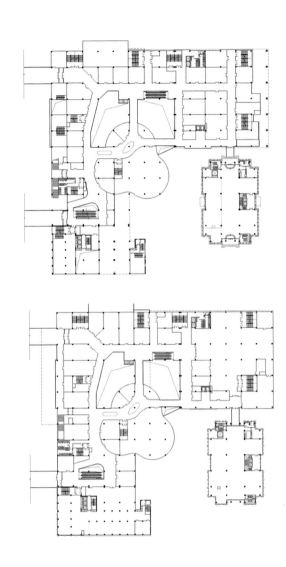

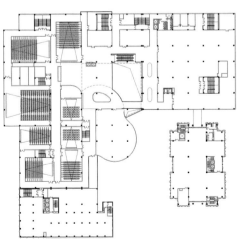

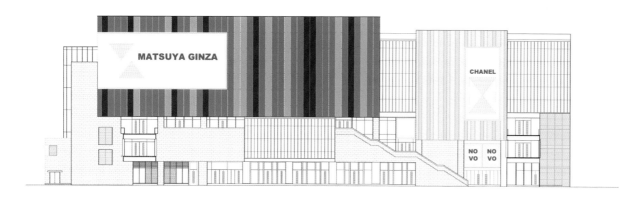

left and right photos and section diagram/实景照和局部详图

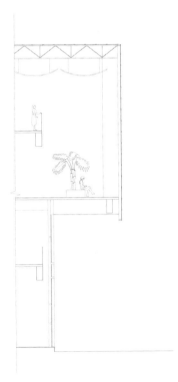
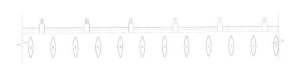

XI'AN ZHENGDA PLAZA

西安正大广场

用地面积	建筑面积	设计时间	建筑类型
39 700平方米	101 718平方米	2010年6月	商业建筑

西安正大广场是西安市高新区倾力打造的大型城市综合体的一期部分，主要业态包括大型超市、品牌旗舰店、百货主题馆、精品商业街、I-max影院、书局和大型餐饮等，更将享誉全球的正大集团旗下的品牌——正大广场带到西安。

建筑造型强调大虚大实，以体量对比形成强烈的视觉冲击力。同时充分考虑与北侧的城市公园的互动性，将自然的主题融入建筑设计，使得建筑人性化、自然化。正大广场和二期270米超高层办公楼建成后，必将带动高新区城市发展及区域人口密度和品质的提升，在高新区乃至整个西咸地区发挥重要的商业和商务中心作用。

Xi'an Zhengda Plaza is phase I project of large urban complex created in high-tech zone of Xi'an City and mainly consists of large supermarket, brand flagship store, shopping theme store, boutique commercial street, I-max cinema, bookstore and large restaurant, etc., even introducing Zhengda Plaza, a brand owned by Zhengda Group which is a world-famous enterprise, to Xi'an.

Building shape shows obvious contrast between virtuality and actuality, using volume contrast to form intensive visual impact. Meanwhile, considerations are made of mutual relation with urban park at north, successfully integrating natural environment into architectural design and realizing humanistic and natural building. After completion, Zhengda Plaza and phase II super-high office building of 270m will certainly accelerate urban development of the high-tech zone and increase population density and improve life quality in this district, playing an important role of commercial and business center in the high-tech zone, even in the whole Xixian District.

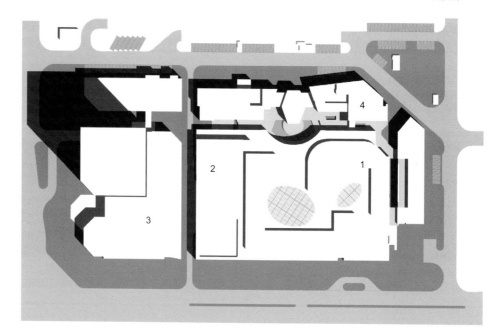

1. Zhengda plaza/正大广场
2. super-high office building/超高层办公楼
3. shopping mall/商场
4. business street/商业街

left   site plan/总平面图
top right   night view of the main entrance/主入口夜景图
bottom right   exterior view/外景图

1. shop/商铺
2. atrium/中庭
3. restaurant/餐饮
4. movie/影院
5. book mall/书城
6. projection booth/放映间
7. path/放映走道
8. air-conditioner room/空调机房

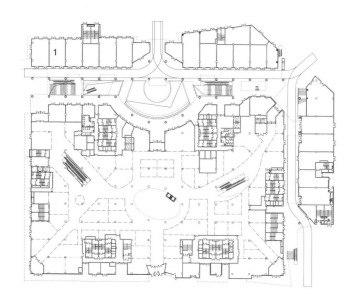
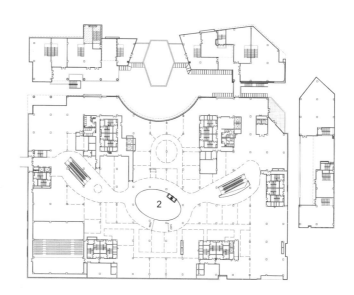
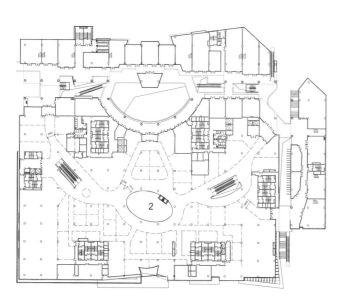
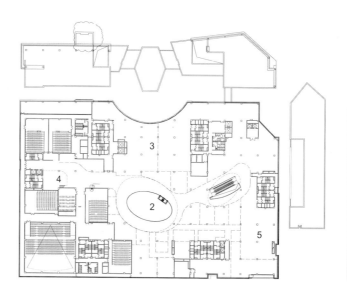

left floor plans/楼层平面图
right exterior view/外景图

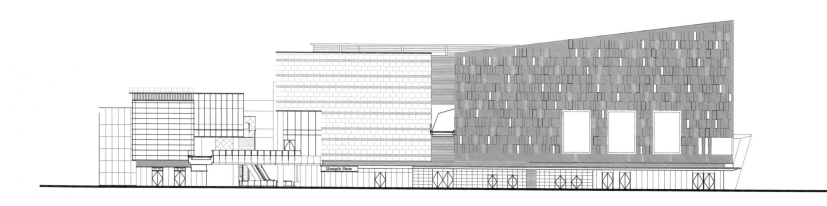
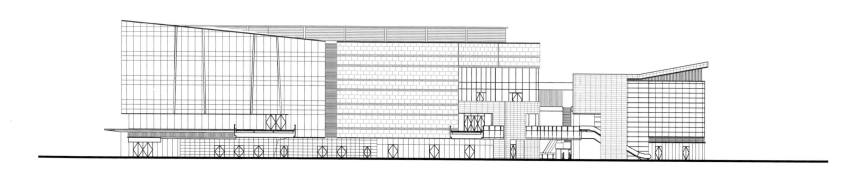

left elevations/立面图
right exterior view/外景图

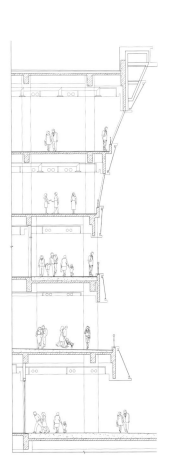

SHANGHAI GREENLAND METRO-PARK PLAZA

上海绿地公园广场

用地面积	建筑面积	设计时间	建筑类型
58 618平方米	22 2969平方米	2009年9月至今	商业+办公

绿地公园广场为地铁上盖物业，位于上海市宝山区顾村镇，向南可环顾整个顾村公园。地铁7号线陆翔路站直接连接到地块东南角，并与地下商业连通。本项目总用地面积为58618平方米。用地性质为商业及办公综合区，主要由5幢高层办公楼及3~4层商业、地下商业及大型地下车库构成。

沿陆翔路规划为大型精品购物中心。设计"十字"及"环路"相结合的流线模式，既有效消化了地块进深，又提供了可分可合的多种店铺组合可能，创造出清晰而又变化多样的购物体验空间。中心处设计一个椭圆形的露天下沉广场，可举办各类展览、节日庆典和其他商业活动。公园广场将成为上海宝山顾村公园板块集大型主题购物、餐饮、娱乐、休闲、文化等一站式综合性"游憩型娱乐休闲购物中心"。

Metro-Park Plaza, located in Gucun Town, Baoshan District, Shanghai City, is built over a subway line and overlooks the whole Gucun Park at south. Luxiang Station of Subway Line 7 is directly connected with the southeast corner of the plot and the underground commercial space. This project, covering a total area of 58,618m², is a commercial and office complex and mainly consists of five high-rise office buildings, three to four commercial floors and underground commercial floor and large underground parking lot.

Large boutique shopping center is planned along the Luxiang Road. Integration of "cross" and "ring" roads not only effectively digests the depth of project site, but also provides flexible and diversified stores and creates clear and various spaces for shopping experiences. An elliptic open sunk square is designed at the central area, which could accommodate different exhibitions, festival celebrations and other commercial activities. The park plaza will become a one-stop "leisure and entertainment shopping center" integrating large theme shopping, catering, entertainment, leisure and culture in Gucun Park of Baoshan District, Shanghai.

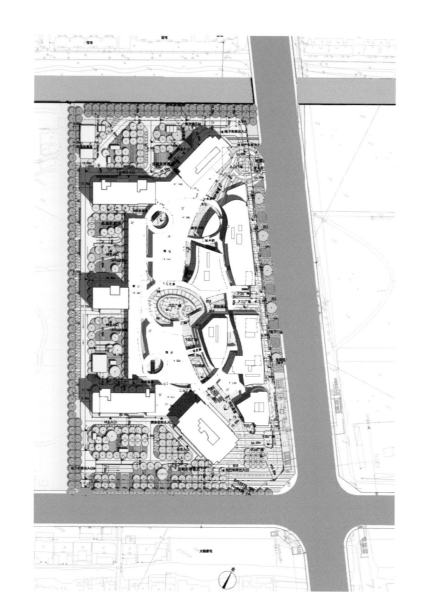

left site plan/总平面图
above bird's eye view/鸟瞰效果图

left and right   commercial inner street perspective/商业内街透视图

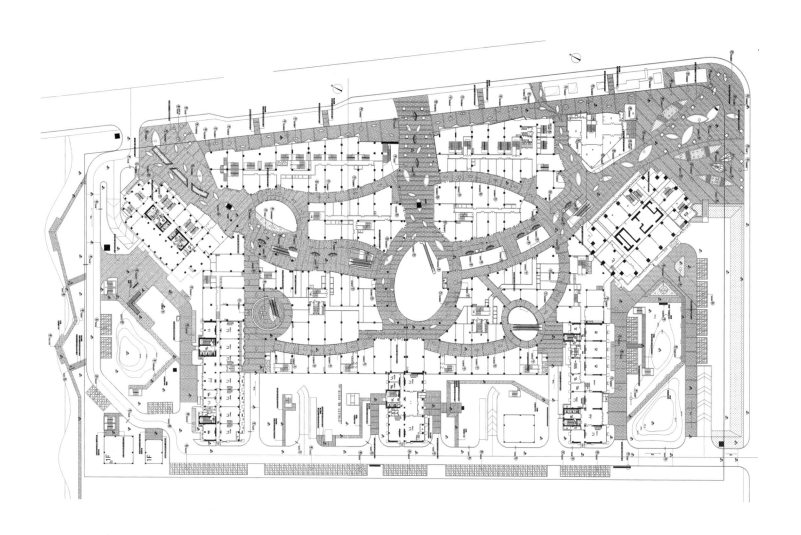
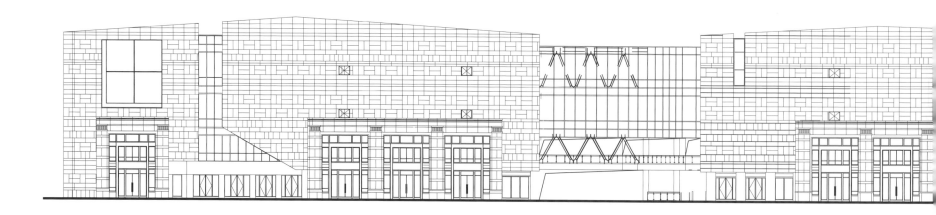

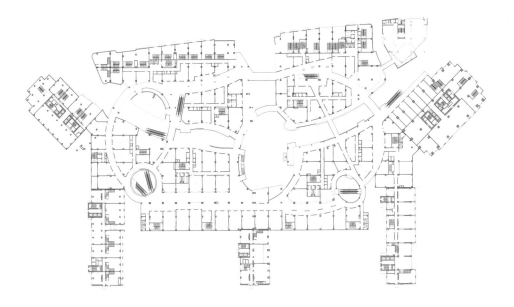

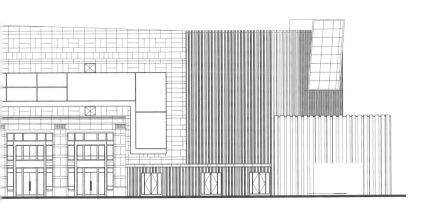

top left   business plan/商业平面图
bottom left   commercial street facades/商业沿街立面图
above   local architectural photo/建筑局部实景图

YANGZHOU YUNHEJI COMPLEX

扬州运河纪地块

用地面积	建筑面积	设计时间	建筑类型
38 528平方米	130 946平方米	2007年8月	酒店办公及商业

扬州运河纪地块北临江阳西路，东临润阳路。由3幢塔楼（甲级办公楼、LOFT办公及福朋酒店）及其裙房、2层购物内街及地下车库构成。目前一期福朋酒店已经建成并投入使用。

规划充分考虑分期实施与分期投入使用的要求，采用清晰理、性的空间规划结构，使建筑群达到了互相依存又相互独立的群体效果。3栋塔楼两两围合，自然形成3个相对独立的商业广场、酒店入口广场和办公入口广场，巧妙地化解了复杂的交通流线。酒店按五星级标准设计与建造，外立面采用简洁平稳的横向线条，顶部和窗体细节刻画入微，给人以精致细腻的感受。材质选用充分尊重扬州当地黑白灰的城市风貌要求，采用含蓄的暖白色铝板与暖灰色石材，形成既符合扬州城市气质，又独具品质感的高档酒店形象。

Yangzhou Yunheji Complex is adjacent to Jiangyang West Road at north and Runyang Road at east. The project consists of three tower buildings (grade-A office building, LOFT office building and Four Points Hotel), podium building, two floors of internal shopping street and underground parking lot. For the time being, Phase I project of Four Points Hotel has been completed and put into service.

The planning gives full considerations to the requirements of construction and completion in phases. Therefore, clear and rational spatial planning structure is adopted to realize mutual support and relative independence among buildings. The three tower buildings form three relatively independent commercial square, hotel entrance square and office building entrance square, skillfully solving the problem of complicated traffic flows. The hotel was designed and constructed according to five-star standard to realize simple and smooth transverse strips on facades and precise details on roof and windows, giving people a feeling of refinement. Building materials well satisfy the local requirements of black, white and grey urban environment in Yangzhou. Connotative warm white aluminium plates and warm grey stones are used to respond to the urban characteristics of Yangzhou and show the high-quality hotel image.

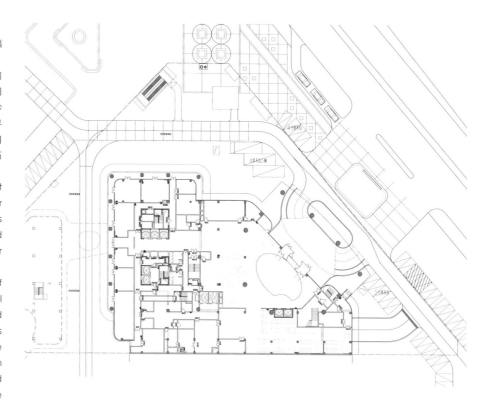

above site plan/总平面图
right exterior view/外景图

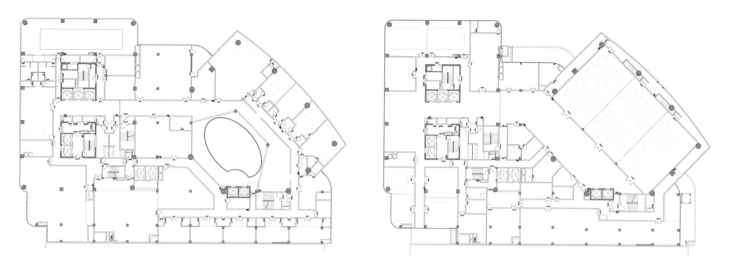

above   plans of Four Points Hotel/福朋酒店各层平面图
below and right   photos/实景照

left   architectural detail photos/建筑细部实景照
right   interior photos/室内实景照

SHANGHAI JUNFENG INTERNATIONAL FORTUNE PLAZA

上海骏丰国际财富广场

用地面积	建筑面积	设计时间	建筑类型
0.827万平方米	5.5万平方米	2006年5月	办公及商业综合体

上海骏丰国际财富广场位于上海市虹口区商业中心地段，大连路与四平路交叉口，M8、M10地铁上盖，比邻和平公园、教育电视台，商业娱乐设施齐全，地理位置优越。地块形状为三角形且用地紧张，周边条件复杂，设计在综合各种外部不利条件的前提下，巧妙布置高层写字楼和商业裙房，创造出与地块高度契合的建筑体量。高层建筑沿大连路在不同高度有不同的退界要求，在60米、80米须有两次退界，设计巧妙结合退界要求形成层层收进的形体特色，并最大限度地利用了土地。建筑风格采用挺拔高耸的Art Deco形式，现已成为周边地区内集商业、餐饮、办公于一体的甲级高档商业办楼。

Shanghai Junfeng International Fortune Plaza, located at commercial center of Hongkou District of Shanghai City and at the crossroad between Dalian Road and Siping Road, is built over metro lines 8 and 10 and near Heping Park and Educational Television Station, enjoying complete recreational facilities and advantageous geographical location. The site is of triangular shape and limited area, having complicated surrounding context. Therefore, the design gives considerations to various external disadvantages and skillfully arranges high-rise office buildings and commercial podium buildings to successfully realize a building volume corresponding to the site elevation. High-rise buildings shall be terraced at different heights along Dalian Road and shall be terraced twice at 60m and 80m heights respectively. The design strictly complies with the planning requirements to realize terraced appearance and to maximize the utilization of land. Buildings are designed in the high Art Deco style, having created a grade-A high-end business building integrating commerce, restaurant and office in surrounding context.

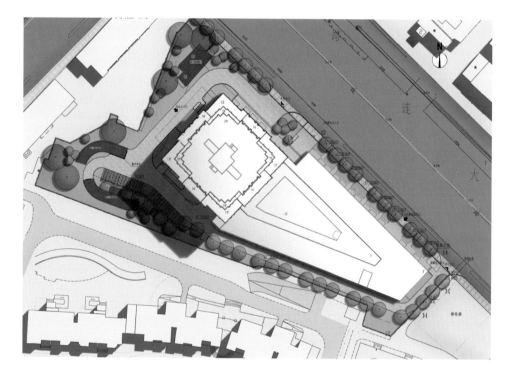

above coloured general plan/彩色总平面图
right photo/实景照

left photo/实景照
below details photo/局部实景照
right elevation&photo/立面图&实景照

BEIJING AIR CITY

北京翼之城

用地面积	建筑面积	设计时间	建筑类型
159 355平方米	152 860平方米	2009年12月	商业金融

北京翼之城位于北京市顺义区顺义新城29街区，隶属北京临空经济开发区，距T3航站楼约3千米，距北京CBD区域45分钟车程。规划顺应地块狭长的形状，构成龙形图案，一气呵成，大气磅礴。由于地处航线之下，建筑外立面不能产生较强的反射光线，因此，办公建筑立面在玻璃幕墙外附加简洁的线条格栅，既有效地杜绝了眩光影响，又通过遮阳提高了建筑的节能效果，形成独特的建筑风格。本案建成后成为了首都机场T3航站楼边上的一道亮丽的风景线。

Beijing Air City, located in block 29 of Shunyi New Town in Shunyi District of Beijing City, belongs to Beijing Airport Economic Development Zone of Beijing City and is about 3km from T3 Terminal Building and 45 minutes driving distance from Beijing CBD. The planning considers the narrow and long site shape and forms a dragon pattern, showing generous and elegant characteristics. Due to the fact that this project is located under the flight course, building facades can not generate relatively strong reflected light. Therefore, simple strip grids are attached over the glass curtain wall on the facades of office building to effectively avoid glare and provide sunshade which improves building energy saving performance and forms unique architectural appearance. After completion, this project has created a beautiful skyline beside T3 terminal Building of Beijing International Capital Airport.

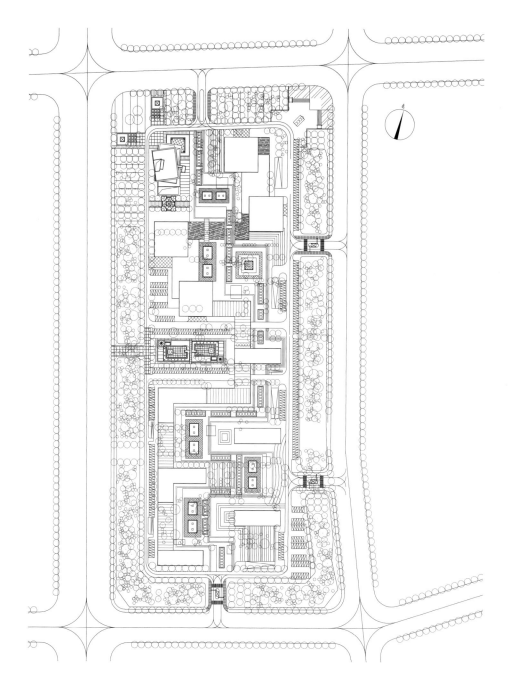

left site plan/总平面图
below bird's eye view/鸟瞰图

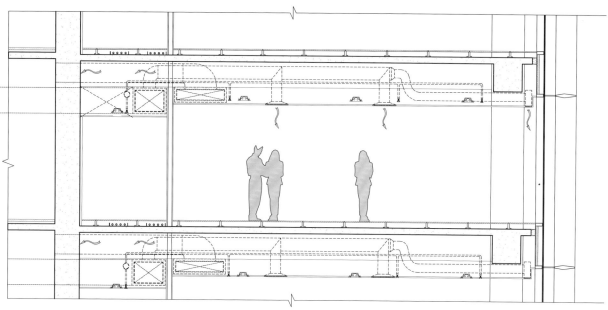

top left and right　exterior view/外景图
bottom left　wall with VAV system section/带VAV系统的墙体剖面图

SHANGHAI CHANGJIU-TAOLE LIFE PLAZA

上海长九陶乐生活广场

用地面积	建筑面积	设计时间	建筑类型
36 245平方米	99 920平方米	2012年4月至今	城市综合体

上海长九陶乐生活广场项目位于上海市嘉定区江桥镇，东至嘉涌路，南至海波路，西至临洮路，北至陇南路，交通便捷。地块周边住区围绕，商业、办公的潜在客户群充足，未来潜力巨大。在项目规划中，关注混时生活（Time-mix Life）、开放空间（Open Space）和人际渗透（Inter personal Penetration）三个概念的融合，共同打造适合创业初期的新兴城市人的办公起航社，同时为周边住区营造具有标志性和吸引力的商业场所。在规划结构上采用外围围合，内部共享庭院的方式，合理规避等级较高的陇南路、临洮路对基地的影响，创造简洁大气、现代动感的外部立面和宁静舒适、灵动丰富的内部空间。在商业区域的布局上，在地块的中央植入了"上海梦"——造型鲜明，具有标志性的视觉中心，像一朵彩云漂浮在商业广场之上；周边以流线边界定义了商业内街，打通了东侧住区到达临洮路、海波路街角的通道，也为建筑群体增添了活力。

Shanghai Changjiutaole Life Plaza, located in Jiangqiao Town of Jiading District, Shanghai City, enjoys convenient traffic system which spreads from Jiayong Road at east to Haibo Road at south, Lintao Road at west and Longnan Road at north. The site is surrounded by residential communities, which will bring sufficient potential customers for commerce and business, making the site have huge development potential. During project planning, the design focused on integration of time-mix life, open space and interpersonal penetration concepts to build an initiative office building for emerging urban people who want to create their business and to create a landmark and appealing commercial space for surrounding residential communities. Peripheral closure and shared court are designed for planning layout to reasonably avoid the influence of Longnan Road and Lintao Road which have higher traffic level on the site and to create simple, generous and modern external facades and peaceful, comfortable and flexible interior spaces. For the layout of commercial areas, "Shanghai Dream" is designed as a visual landmark at the central area. It is just like a cloud close floating over the commercial plaza; internal commercial street is defined by surrounding streamline boundary, forming a passage from the residential community at east to the corner of Lintao Road and Haibo Road and infusing more vigor into the building cluster.

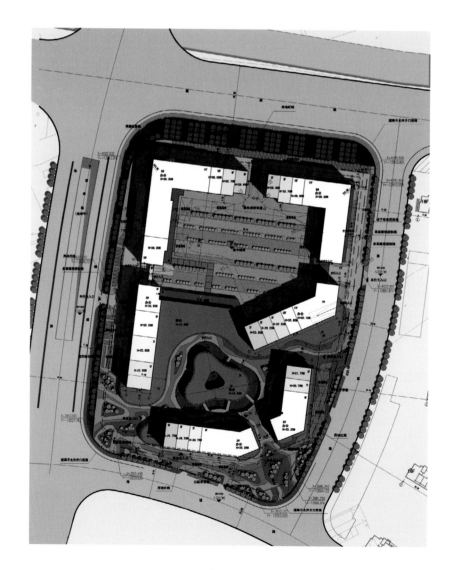

left   site plan/总平面图
below   bird's eye view/鸟瞰图

above and below   exterior view/外景图
right   business perspective/商业透视图

IMPRESSION OF YING LAKE

英湖印象

用地面积	建筑面积	设计时间	建筑类型
5.80万平方米	9.94万平方米	2007年至2008年6月	住宅

本项目位于吉林省长春市西环城路以东、规划路以南、规划用地以西、景阳大路以北。项目紧靠景阳大路和西环城路两条连接市区的主要干道，而且景阳大路作为贯穿长春南北市区的交通纽带被长春市民称做"第二条人民大街"，西环城路是长春市区外环干道，也是市区进入一汽集团厂区的主要通道。

Impression of Ying Lake, located in Changchun City of Jilin Province, spreads to West Ring Road at east, Guihua Road at south, planned site at west and Jingyang Road at north. The project is adjacent to Jingyang Road and West Ring Road, which are two trunk roads connecting the downtown area. Jingyang Road, a main transportation link, passes through urban hub from south to north and is called by Changchun citizens as "the Second Renmin Avenue", while West Ring Road is an external ring trunk road in Changchun and it is also a main passage connecting the downtown area to the plant of FAW Group.

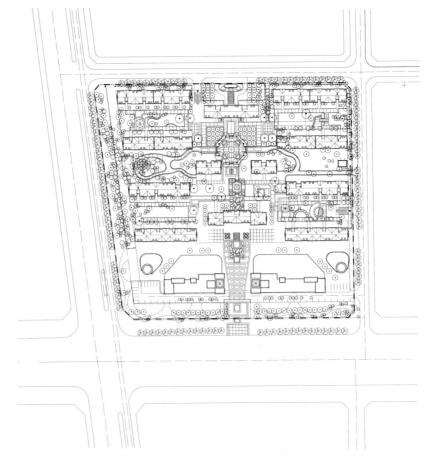

above   site plan/总平面图
right   architectural photo/建筑实景照图

top left   local architectural photo/建筑局部照
bottom left   architectural photo/建筑实景照
right   architectural photo/建筑实景照

YANCHENG GREENLAND BUSINESS CITY

盐城绿地商务城

用地面积	建筑面积	设计时间	建筑类型
57.1万平方米	131.5万平方米	2011年至2011年10月	商业+办公+住宅

绿地商务城位于盐城市东部亭湖新区的核心区域，是服务于盐城主城区160万人口，集商业中心、休闲娱乐、商贸办公、金融服务、文化会展、餐饮服务、居住生活等多种现代城市功能于一体的超大型CBD+CLD综合社区。绿地商务城的建成确立了其在亭湖区的核心地位。

Greenland Business City, located in the core area of Tinghu New District at eastern part of Yancheng City, is a super-large CBD+CLD general community integrating multiple modern urban functions, such as commercial center, leisure and entertainment, buisiness and trade office, financial service, cultural exhibition, restaurant and residential building, serving for 1.6 million citizens in downtown area of Yancheng City. After completion, this project has become a core landmark at Tinghu District.

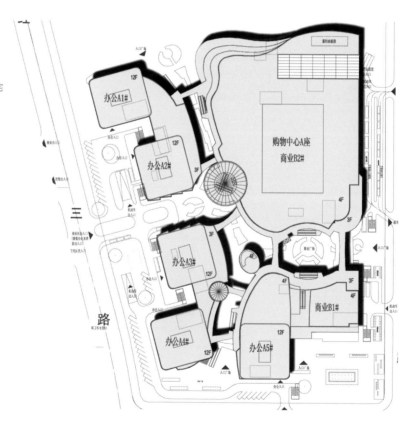

above site plan/总平面图
right business perspective/商业透视图

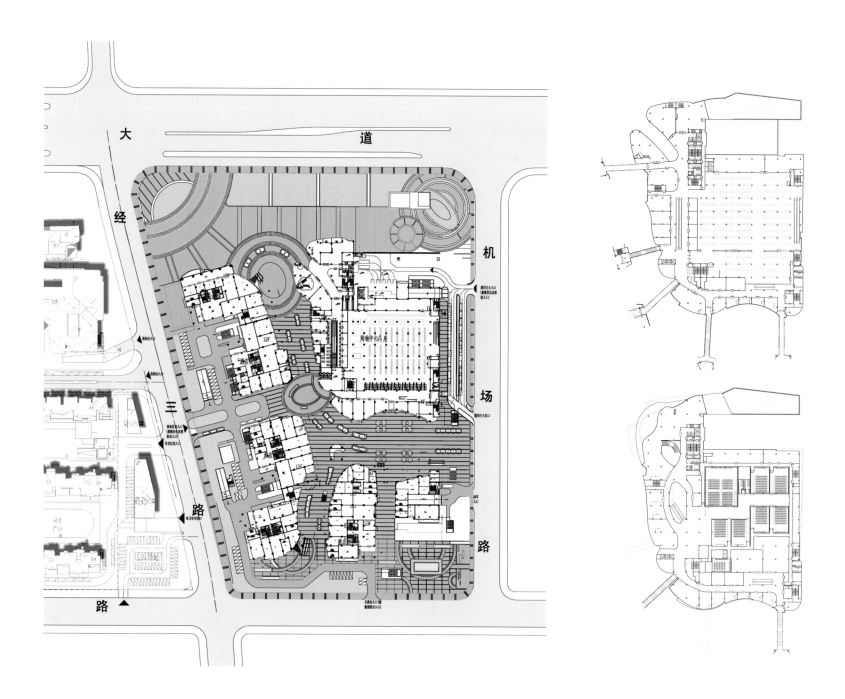

left commercial 1st floor plan/商业一层平面图
right business perspective/商业透视图

XUZHOU HIGH-SPEED RAILWAY STATION PROJECT PLOT I

徐州高铁站前项目I地块

用地面积	建筑面积	设计时间	建筑类型
36 551平方米	147 988平方米	2010年5月	办公及商业综合体

徐州高铁站前项目I地块北临高铁北路，东临站场路，南临狼山路，西临站前路。规划以完整和谐的商业与办公商务相结合的思想为指导，同时体现清晰理性的空间规划结构，达到与城市空间的合理衔接以及地块内部各功能区域的完美结合。整体建筑形象由一栋创意办公、一栋办公和下部商业裙房三大部分组成，形成高低错落的外部空间形象。整个建筑形态舒展、均衡，彰显大气、稳重的气派。

Xuzhou High-speed Railway Station Project Plot I is adjacent to Gaotie North Road at north, Zhanchang Road at east, Langshan Road at south and Zhanqian Road at west. The planning is based on the concept of completed and harmonious integration of commercial and business functions, as well as clear and reasonable spatial planning structure, so as to realize reasonable connection with urban space and perfect combination among different functional areas on the plot. The overall building is composed of one creative office building, one office building and commercial podium buildings below, forming diversified external spaces. The building shape shows smooth, balanced, generous and solemn characteristics.

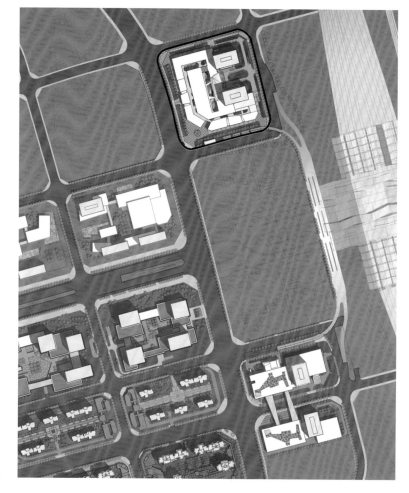

right project site/项目基地
far right exterior view/外景图

left   1st floor plan, 2nd floor plan/一层平面图，二层平面图
top right   exterior view/外景图

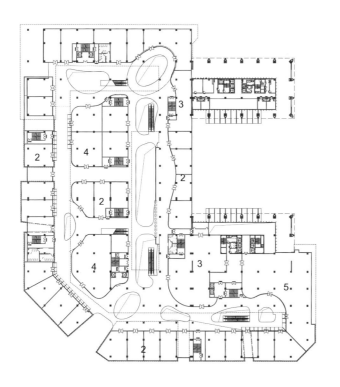

1. hall/大堂
2. shop/商铺
3. anchor store/主力店
4. boutique shop/精品商铺
5. theme restaurant/主题餐饮

1. entrance detail/门头节点
2. curtain-wall unit detail/幕墙单元格详解
3. annex's elevation detail/裙房外立面详解

1

2

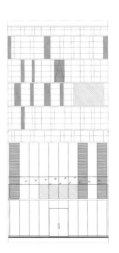

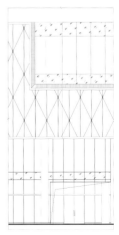
3

below axis elevations/轴立面图
right exterior view/外景图

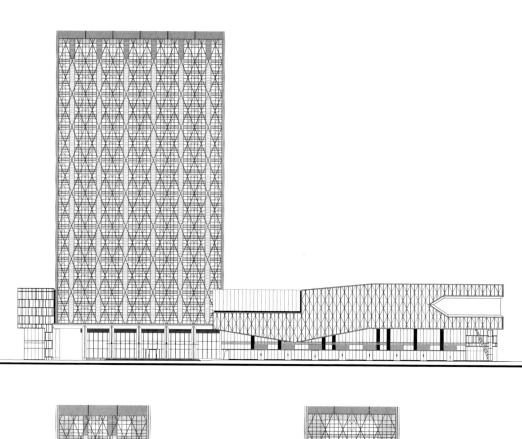

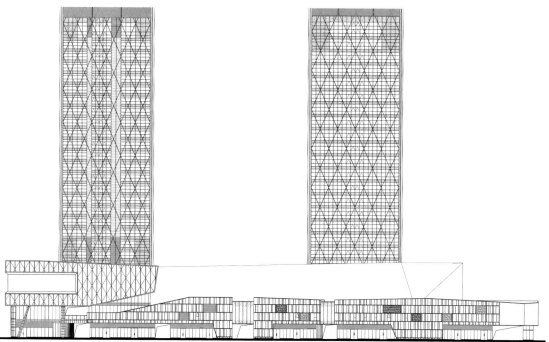

XUZHOU HIGH-SPEED RAILWAY STATION PROJECT PLOT J

徐州高铁站前项目J地块

用地面积	建筑面积	设计时间	建筑类型
18 153平方米	57 660平方米	2010年5月	酒店及商业综合体

徐州高铁站前项目J地块北临珠山路，东临站场路，南临高铁南路，西临站前路。功能定位为商业及酒店综合区，土地开发强度相对较高。地块规划以"两区、一轴、多节点"展开，形成完整和谐的商业与酒店商务功能结合体，体现清晰理性的空间规划结构，达到了与城市空间的合理衔接以及地块内部各功能区域的完美结合。建筑由假日酒店塔楼及其商业裙房以及西侧商业区三个大的部分组成。塔楼和裙房形成良好的空间对比和景深感，裙房采用连续的立面处理方式，使得整个建筑形态舒展、均衡，彰显大气、稳重的气派。

Xuzhou High-Speed Railway Station Project Plot J is adjacent to Zhushan Road at north, Zhanchang Road at east, Gaotie South Road at south and Zhanqian Road at west. Functional orientation of this project is integration of commerce and hotel, which requires relatively high development standard. The planning is based on the principle of "two zones, one axis and multiple points" to form a completed and harmonious integration of commerce and hotel business functions, showing clear and rational spatial planning layout and realizing reasonable connection with urban space and perfect links among various functional areas on this plot. Buildings consist of three parts, including holiday hotel tower building, commercial podium building and commercial area at west. The tower building and podium building have favorable spatial contrast and perception of depth, while podium building adopts consecutive facade treatment to produce extended, balanced, generous and solemn appearance.

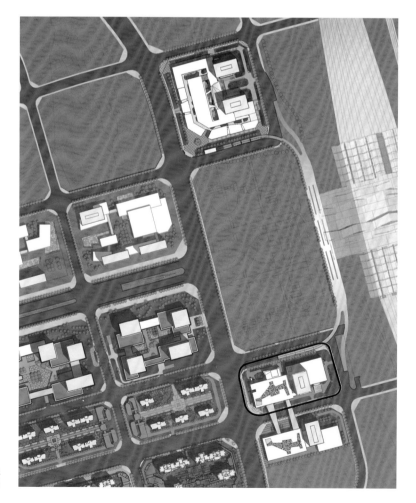

right project site/项目基地
far right exterior view/外景图

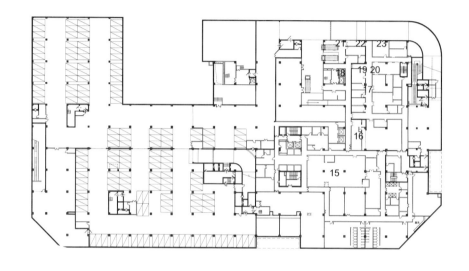
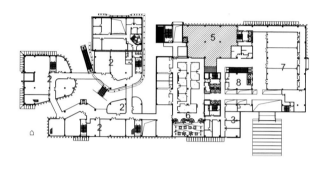
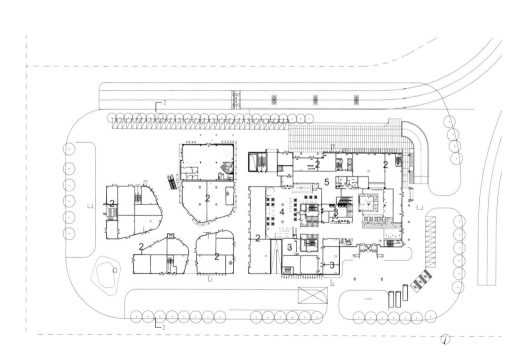
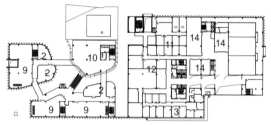
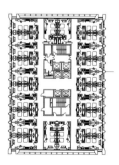

left   plans/各层平面图
right exterior view/外景图

1. hall/大堂
2. shop/商铺
3. office/办公
4. 24 hours restaurant/全日餐厅
5. kitchen/厨房
6. Chinese restaurant/中餐厅
7. banquet room/宴会厅
8. VIP reception room/VIP接待室
9. anchor store/主力店
10. terrace/露台
11. KTV
12. fitness room/健身房
13. SPA
14. meeting room/会议室
15. wash room/洗衣房
16. staff kitchen/员工厨房
17. bakery room/烘焙间
18. veg prep/蔬菜加工间
19. meat prep/肉类加工间
20. seafood prep/海鲜加工间
21. wine storage/酒水仓库
22. dairy chiller/奶制品冷库
23. dry store/干货库

left　exterior view/外景图
below　south and north elevations/南立面、北立面

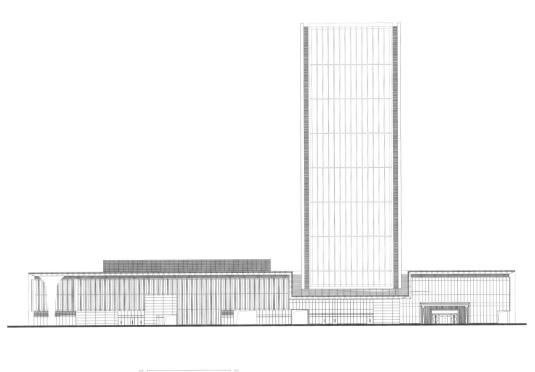
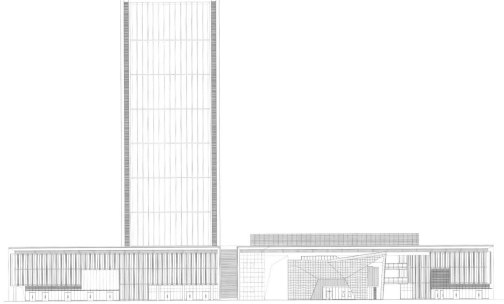

3 文化教育建筑
Culture and Education Building

创新是建筑师永恒的追求。文化教育建筑的特定属性为精神层面的空间创作提供了充分的自由度。密斯说："当两块砖相遇即产生了建筑。"这是怎样精彩的相遇啊？冰冷的构造不可能产生激动人心的建筑，中国古代画论曰"情为意本，意在笔先"，只有建筑师将材料进行有情有义的倾心编织，才能达到密斯终生追求的"技术升华为艺术"的至高境界，创作出动人心灵的文化空间。

Innovation is a constant pursuit of architects. Specific characteristics of cultural and educational building provide sufficient freedom for spatial creation in spiritual aspect. Mies said, "It is the encounter of two bricks that generates the building". How splendid encounter it is? Inanimate structure never generates outstanding building. Chinese ancient drawing theory believes that "fervor is the essence of idea and idea foregoes the brush pencil".Architects can not realize "technology upgrading to art" which is pursued by Mies in all of his life and create charming cultural space without paying their enthusiasm into building materials.

SHENYANG QIPANSHAN NATIONAL CONFERENCE CENTER

沈阳棋盘山国家会议中心

用地面积	建筑面积	设计时间	建筑类型
36 611.6平方米	34 850平方米	2011年1月	会议中心

项目位于棋盘山风景区内，处于沈阳市东北部。西、南为沈阳市城区，距沈阳市中心约20千米。

整体布局利用优越的区域环境，将山水建筑融为一体，打造"自然、生态、健康、休闲的"国家级会议接待中心。造型汲取中国古典建筑的元素及其比例，承载关于中式汉唐的记忆。汉唐文化的大气与厚重通过方正的造型母题、沉静的色调、舒展平远的屋顶及细部精巧的线条呈现，渗透着汉唐宫殿的气魄，宏伟而优雅。沿着自然与历史的轨迹将汉唐开放的社会意识融入其中，并将这种气质与功能完美结合，力求实现一种梦归唐朝的特殊氛围。

Qipanshan National Conference Center, located in Qipan Mountain Resort at northeastern part of Shenyang City, faces the downtown area of Shenyang at west and south and is about 20km far from the urban center. The general layout utilizes the advantageous context environment to integrate the building into mountain environment and waterscape, successfully creating a national conference reception center focusing on natural, ecological, healthy and recreational characteristics. Building shape makes references to Chinese ancient architectural elements and proportion, recalling people's memory of Chinese architectural culture of Han and Tang Dynasties. The cultures of Han and Tang Dynasties are generous and profound, showing palatial and elegant features of royal buildings through regularly square shape, peaceful color, smooth roof and delicate details. The open social consciousness of Han and Tang Dynasties is infused into building according to natural and historical conditions, the elegant appearance is perfectly integrated with architectural functions to reproduce a special atmosphere of dreaming in ancient dynasties.

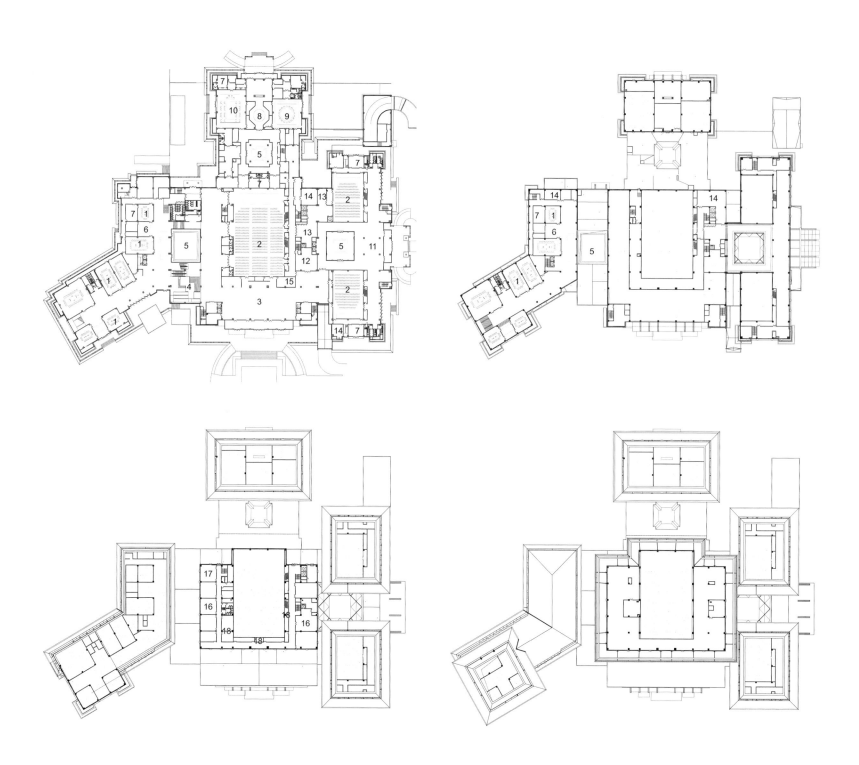

left plans/各层平面图
right bird's eye view/鸟瞰图

1. meeting room/会议室
2. multifunctional hall/多功能厅
3. main hall/主大堂
4. atrium/中庭
5. landscape plaza/室外庭院
6. refreshment room/茶歇室
7. guest office/贵宾室
8. central hall/中厅
9. banquet room/宴会厅
10. news studio/新闻发布厅
11. subordinate hall/次大堂
12. hall of administration/行政大厅
13. cater room/备餐室
14. furniture storage/家具储藏
15. business guest office/商务接待
16. administration guest office/行政接待室
17. media lounge/记者休息室
18. facility control room/控制室

above elevations/立面图
below exterior view/外景图

LIAONING CULTURE PLAZA

辽宁省文化广场

用地面积	建筑面积	设计时间	建筑类型
30.46万平方米	35.3万平方米	2011年3月	文化类大型公共建筑

辽宁省文化广场位于沈阳市新规划的浑南新城中心区域。规划将图书馆、科技馆、博物馆、档案馆四馆围绕公共性市民广场统一布局，并以"绿色魔毯"理念，通过提升、下沉、分割、覆盖等手法，将四馆融为一体，有效提高广场的群众参与性。建筑设计引入参数化技术及GRFC等新材料，营造出具有浓郁时代感的生态型城市文化景观广场。

Liaoning Culture Plaza is located at the central area of newly-planned Hunnan New Town in Shenyang City. The planning scheme uniformly sets the four pavilions, including library, scientific and technological pavilion, museum and archives, around a public square and, based on the concept of "green magic carpet", integrates the four pavilions through lifting, sinking, partition and covering measures, so as to effectively improve people's experiences on the square. Parametrization technique and GRFC new material are introduced into architectural design to create ecological urban cultural landscape square with modern features.

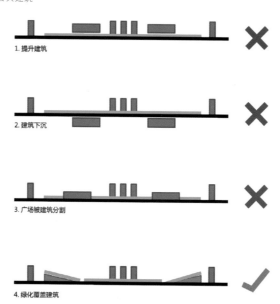

1. 提升建筑 ✗
2. 建筑下沉 ✗
3. 广场被建筑分割 ✗
4. 绿化覆盖建筑 ✓

right combination of site and building/场地与建筑融合
far right model of four pavilions and plaza/四馆与广场模型

below rendering/效果图
right bird's eye view of the museum and perspecitive of the library/博物馆鸟瞰效果图和图书馆透视效果图

图书馆

CHINA NATIONAL ACADEMY OF PAINTING

中国国家画院

用地面积
58,175平方米

建筑面积
10,375 平方米

设计时间
2011年5月

建筑类型
文化娱乐建筑等

中国国家画院创作基地，坐落在天津市蓟县5A级国家名胜风景区。创作基地主要是提供画院艺术家进行艺术创作和展览使用，一共有11栋楼；4栋全国最大的地下百米覆土大画室、6栋个人工作室和1栋公共创作院。风格属于具有"诗性"美的中国古典园林建筑。风格体现在四方面：立面——"墙"、内部——"院"、色彩——"黑白灰"、细部——"提炼"。画院建筑与自然完全融合，传承古代园林的精髓。

China National Academy of Painting (CNAP) base is located at the AAAAA national scenery resort in Jixian County of Tianjin. The base is mainly used by artists of CNAP for artistic creation and exhbiition of their works. The project consists of 11 buildings, including four biggest earth-covered painting studios built at about 100m underground in China, six private studio and one public studio. Architectural style of these buildings belongs to Chinese classical garden architecture with "poetic" aesthetics, which is mainly demonstrated in the following four aspects: facade - "wall", interior - "court", color - "black, white and grey" and detail - "refined". These buildings are well integrated into the natural environment and show the essence of Chinese ancient garden architecture.

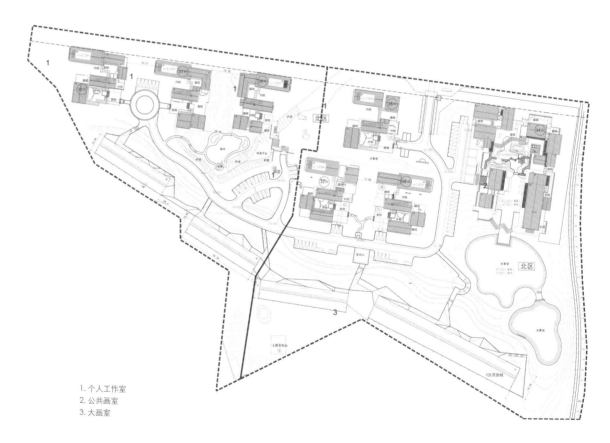

1. 个人工作室
2. 公共画室
3. 大画室

left site plan/总平面图
above photo/建成实景照
below private studio elevations/个人工作室立面图

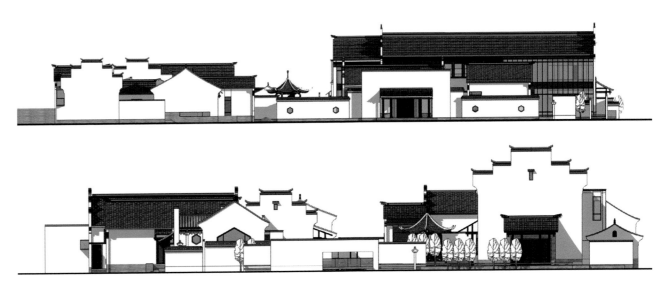

below interior view/内景效果图
middle public studio site plan/公共画室平面图
right detail photos/细部实景照

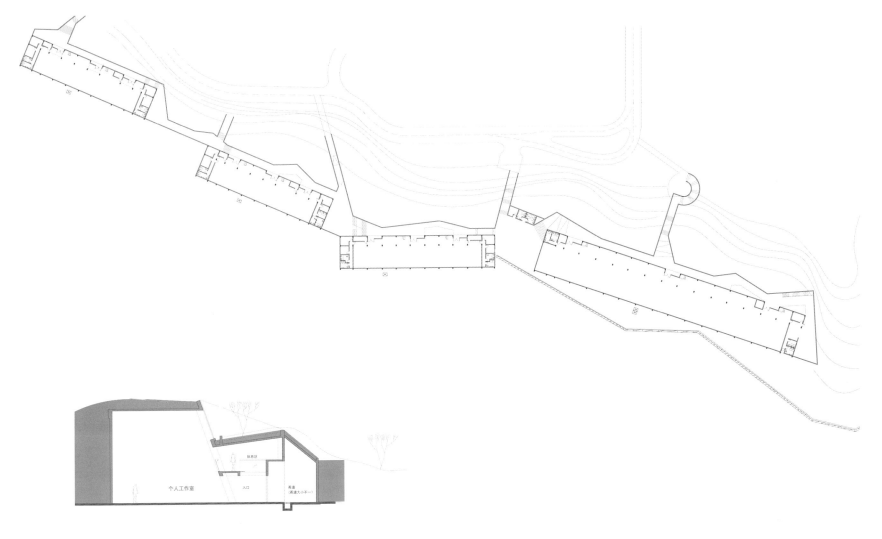

left   interior photo/内部实景照
above big studio site plan and elevations/大画室总平面图和立面图
right interior photo/内部实景照

SALES BVILDING OF BEIJING AIR CITY

北京翼之城售楼处

用地面积	建筑面积	设计时间	建筑类型
约3 000平方米	约2 000平方米	2009年12月	售楼处

北京翼之城售楼处位于首都国际机场东南侧,总建筑面积约2,000平方米。建筑形态轻盈现代,富有冲击力,以大尺度的悬挑和扭转形了极富张力的造型。同时设计充分尊重北京的历史传统,在设计之初把对北京四合院的钟情融合到设计之中,在建筑设计中将"院子"的概念贯彻始终,成为统一全局的元素。建筑师将每一层理解为一个场所,传统院落形式加以现代元素的重构,在每一层都塑造了具有传统意味的现代庭院空间。每层院子都有其独特的主题、鲜明的个性。穿梭其中时心理会产生戏剧性的变化,进而对售楼产生积极的影响。建筑在大量使用现代建筑常用的铝合金型材、穿孔钢板、玻璃的同时,又添加了黑与白、明与暗的对比,凸显其沉稳一面。整体建筑前卫现代又不失传统韵味、彰显个性又不失成熟稳重。

Sales Building of Beijing Air City is located at the southeast of Beijing Capital International Airport and covers a total building area of about 2,000m². The building has vigorous and modern features and produces strong visual impact, utilizing large-scale cantilever and torsion to form an unique appearance. Meanwhile, the design gives full respect to the historical culture of Beijing. Features of Beijing courtyard building are infused into the architectural design and "courtyard" concept penetrates through the whole design process and becomes an element unifying the general layout. Architects design each floor into a location which consists of traditional courtyard and modern elements. Therefore, modern courtyard with traditional culture is created on each floor. Moreover, unique theme and identified features are designed for each courtyard. People's mind may change dramatically while walking through the courtyard, as it actively promotes people's willing to buy one. Besides ordinary building materials, such as aluminium alloy products, perforated steel plate and glass, the building also adds obvious contrast between black and white, as well as light and dark, showing the solemn appearance of building. Overall appearance of the building realizes a perfect balance between modern and traditional culture, personality and solemnity.

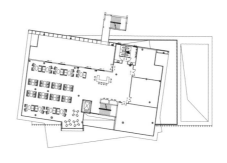

right plans/各层平面图
far right exterior view/外景图

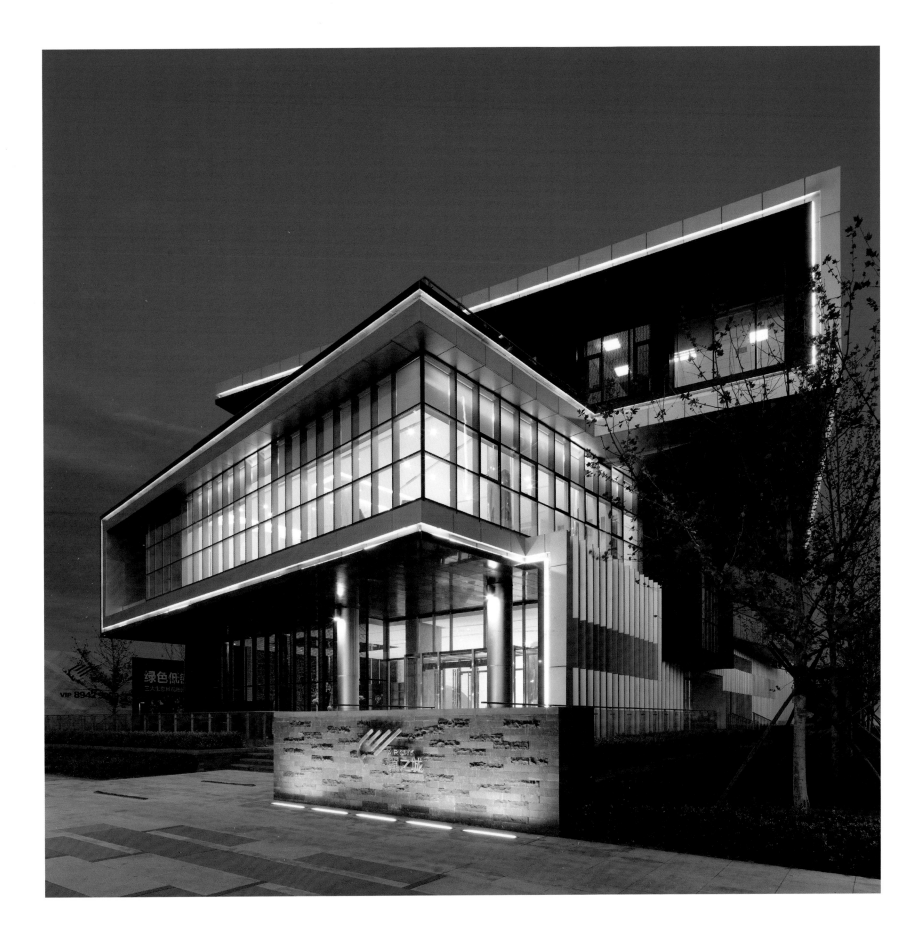

left and right exterior spaces photos/外部空间实景照

left interior spaces photos/内部空间实景照
right node section/节点详图

CHANGTAI XIDONG INTERNATIONAL CITY CHAMBER

长泰锡东国际城会所

用地面积	建筑面积	设计时间	建筑类型
100 603平方米	94 926平方米	2011年至今	办公、商业

本案位于无锡V-park中的核心区域，是展示无锡形象的重要窗口。项目北临先锋大道，南濒东安大道，西起新竹路，东至坊城大道。社区是较大规模的集中商务办公区，将重点吸引国内外公司来无锡设立总部、研发中心、运营中心、销售中心等。本会所是社区最先呈现的建筑单体，兼具为整个商务区服务配套的功能。平面设计采用内外不同层标高进入的设计手法，创造出丰富有趣的地形起伏效果。建筑外观采用典型的艺术装饰派（Art Deco）风格，建筑细节精致而富有逻辑。特别是在建筑入口、塔楼、女儿墙顶部等重点部位集中展现建筑的细腻感和品质感，建成后成为同类风格建筑的典范。

This project is located at the core area in V-park of Wuxi City and is an important portal for exhibition of urban image of Wuxi. The project is adjacent to Xianfeng boulevard at north, Dongan boulevard at south, Xinzhu Road at west and Fangcheng boulevard at east. The block is a large-scale concentrated business zone focusing on attracting domestic and foreign companies to establish their headquarters, research and development center, operation center and distribution center in Wuxi. This chamber is the first single building constructed on the block and it also provides supported services for the whole business zone. Plane design adopts different elevations between internal and external spaces, creating interesting undulant effect. Building appearance is of typical Art Deco style, showing refined and reasonable archiectural details. Especially architectural delicacy and quality are fully demonstrated at entrance, tower building and top of parapet. After completion, it has become a domenstration project for similar buildings.

below   site plan/总平面图
right   exterior view/外景图

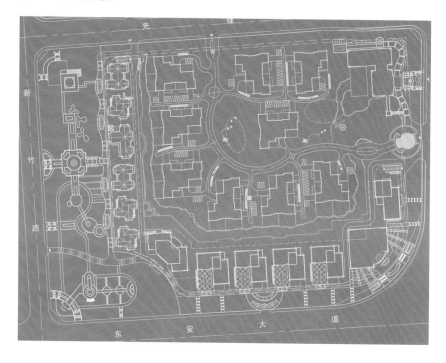

left   architectural photo/建筑实景
above   exterior view/外景图
bottom right   elevations/立面图

left　exterior photo/外部实景照
right　interior photo/内部实景照

TIANJIN DRAGON VALLEY CULTURAL CITY CHAMBER

天津盘龙谷文化城会所

用地面积	建筑面积	设计时间	建筑类型
2 500平方米	1 800平方米	2009年12月	商业金融

盘龙谷文化城会所兼作天津市蓟县盘龙谷星梦工坊的售楼处，建筑师试图去探寻售楼处的本质，从世俗和艺术两种层面上刻画建筑。在这里有两条流线，一条是在人为引导下，经过四个主要场景最终以达成销售为目的的流线，这条流线的空间场所曲折变化，为营造销售氛围而造。另一条流线是自由的，带有文艺气质的，它游走于景观和建筑之间，有着相当的不确定性，建筑师通过景观与建筑的结合塑造了无数变化的场所。建筑参与景观构成，景观参与建筑流线，二者共同塑造建筑空间。

Dragon Valley Cultural City Chamber is also used as sales building in Tianjin City. Architects try to find the essential function of sales building and define the building in two viewpoints: public and artistic. There are two streamlines, one of which uses artifical elements to guide customers passing through four main locations and to sell the building, and is curved according to spatial conditions and is created to generate a selling atmosphere; another streamline possesses freedom and cultural characteristics and stands between landscape and building, having high uncertainty. Architects create numerous variable locations through integration of landscape and building. Therefore, building also becomes a landscape element and landscape defines building profile. They jointly create the architectural space.

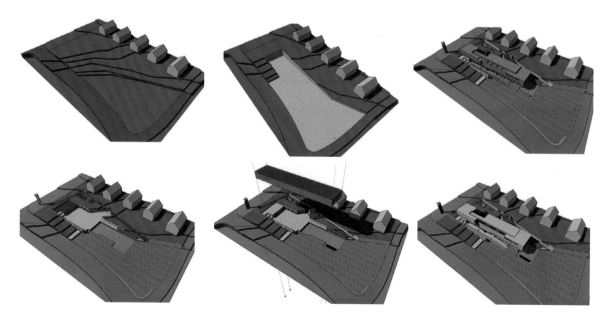

left concept diagram/概念图解
right exterior view/外景图

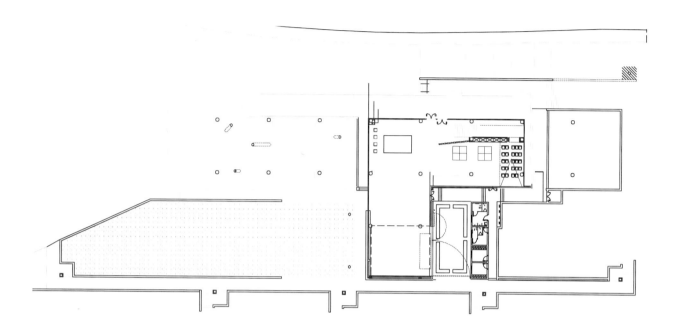
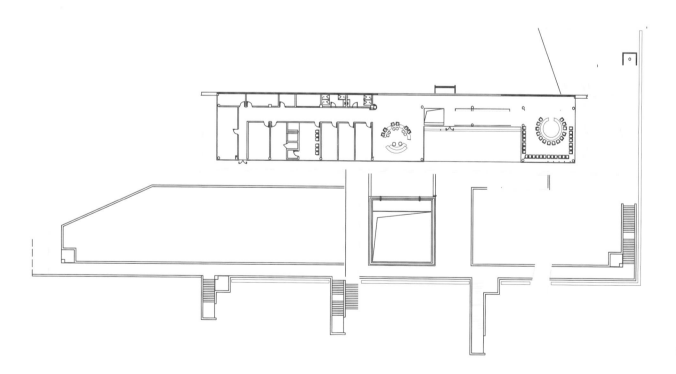

left plans/各层平面图
right exterior spaces photos/外部空间实景照

left elevations/立面图
right exterior spaces photos/外部空间实景照

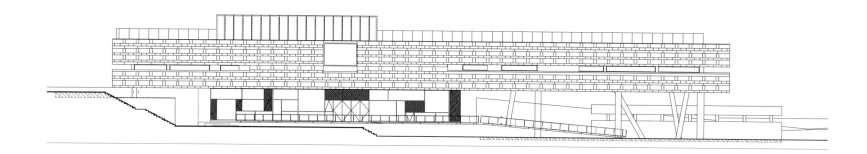

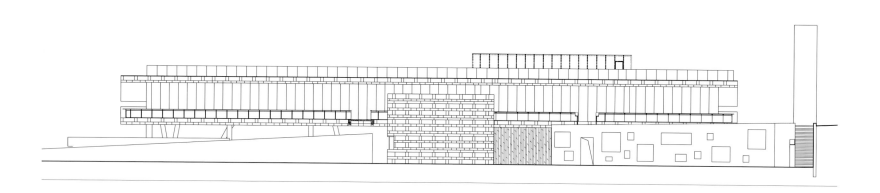

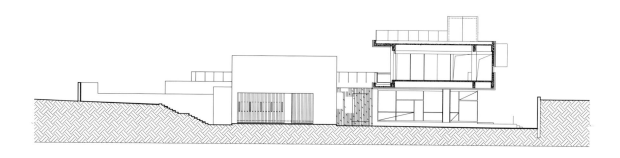

CHANGTAI DIANSHAN LAKE GARDEN CHAMBER

长泰淀湖观园会所

用地面积	建筑面积	设计时间	建筑类型
9 000平方米	5 600平方米	2010年	商业会所

长泰淀湖观园会所位于昆山淀山湖畔，拥有世外桃源般的自然生态环境，作为一个五星级度假酒店的配套部分。建筑设计强调与大自然的融合，建筑如同展开双臂拥抱淀山湖，拥抱自然。内部功能包括大堂、餐饮、健身、室内外游泳池及SPA等，规划采用多院落组合的庭院式布局。轴线关系无处不在，院落空间层层递进。设计选用粗犷的文化石、质朴的涂料和亲人的木材，营造出一个低调奢华、原汁原味的托斯卡纳风情建筑。

Changtai Dianshan Lake Garden Chamber, located by Dianshan Lake in Kunshan City, has beautiful natural and ecological environment and is a supported project of a five-star holiday hotel which will be planned in the near future. Architectural design emphasizes on the integration with natural landscape to realize an architectural shape opening its arms to hug the Dianshan Lake and natural environment. Internal functions of the chamber consist of lobby, restaurant, fitness, indoor and outdoor swimming pools and SPA and the general planning realizes an integration of multiple courtyeards. Axial relation is realized everywhere and diversified courtyard spaces are successfully created. Rough cultural stone, simple coating and moderate timber are used to create a true Tuscany-style building with modest luxury.

below site plan and frist floor plan/总平面图和一层平面图
right bird's eye view/整体鸟瞰图

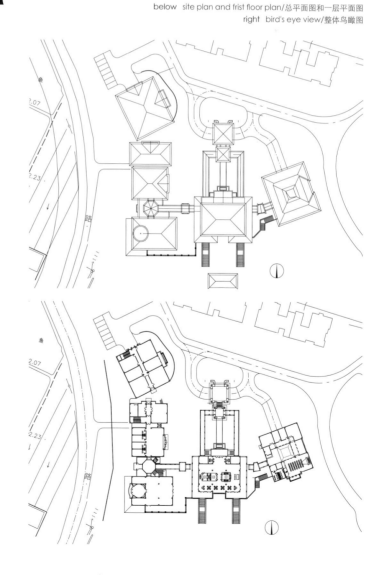

below facade panoramagram/立面全景

above facade panorama/立面全景
right parking space/落车亭
far right inner court entrance/内庭入口

below progressive courtyard/院落递进
right architectural space and detail/建筑空间及细部

left and right　clubhouse indoor setting/会所内景

2011 XI'AN WORLD GARDEN EXPO EXHIBITION PAVILION

2011西安世界园艺博览会绿地生态展示馆

用地面积	建筑面积	设计时间	建筑类型
1 029平方米	1 257平方米	2009年7月至2010年10月	展馆

项目位于"2011西安世界园艺博览会"的企业展园中。园区用地面积1029平方米，建筑占地约500平方米，总建筑面积1,257平方米。基地南北有1.7米的高差，建筑将主入口设计在二层，由园区东南角随一条环形坡道进入建筑。建筑造型具象地将建筑设计成一颗类似被人类砍伐而残留的树根，用超现代的设计表达手法契合博览会的主题。建筑平面是循环圆形，"0"的平面形式表达了我们的核心设计理念"零建筑"——"零介入、零能耗、零排放"。

Greenland Ecological Exhibition Pavilion is located in the enterprise exhibition garden of 2011 Xi'an World Garden Expo. The garden covers a total land area of 1,029m² and the pavilion covers an area of about 500m² and a total building area of 1,257m². The site has a 1.7m elevation different between south and north, so main entrance to the pavilion is designed on the second floor along a ring ramp at the southeast corner of the garden. The building shape is designed as residual root of a tree, which is a ultramodern design method to show the expo theme. Building plane is a circular ring like the numeral "0", which expresses the core design concept of "zero building", namely "zero intervention, zero energy consumption and zero discharge".

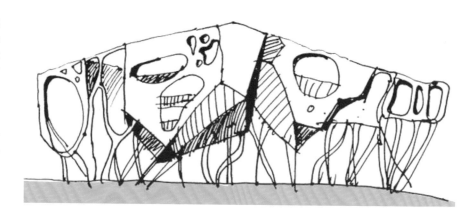

above hand painted intention/手绘意向
right bird's eye view/鸟瞰效果图

left and right architectual photos/建筑实景照

BEIJING SISLEY CHATEAU CHAMBER

北京西斯莱公馆

用地面积	建筑面积	设计时间	建筑类型
163 983平方米	401 500.7 万平方米	2009年–2010年12月	住宅

  北京西斯莱公馆是绿地集团进入北京市场的第一个项目，而本会所是呈现出来的第一个建筑形象。因此，从设计之初就明确了建筑的定位取向：既能体现绿地集团作为上海企业的地域特点，又能体现楼盘高端的定位。设计师汲取最典型的上海外滩建筑群片段意向，严格遵守古典主义建筑构图原则和细部控制法则，创造出具备典型海派特点的建筑形象。建筑的细部控制耗时较长，花饰均取型于上海外滩典型建筑，通过建筑师设计放样并由厂家手工打造完成。会所功能包含售楼处、健身中心、高级餐厅等，特别是在顶层的铜皮屋顶下，设计了一个可以仰望星空的游泳池，成为一个别具特色的场所。

Beijing Sisley Chateau is the first project created by Greenland Group in Beijing and this chamber is the first building to be created. Therefore, Greenland has determined its architectural orientation from the very beginning of design process, namely to show Greenland Group's regional characteristics as a famous Shanghai enterprise and to express the high-end features of the project. Designers refer to the most typical concept of building cluster at The Bund of Shanghai and strictly follow the principle of classical architectural structure and detail control rules to create a building image showing typical Shanghai-style architectural characteristics. Architectural detail control takes a long design period, because all patterns are abstracted from typical buildings at The Bund of Shanghai and are designed by architects before being manually produced by manufacturer. Function of the chamber consists of sales building, fitness center and high-quality restaurant. Especially, an open swimming pool is designed under the copper roof on the top floor, becoming an identified location in the chamber.

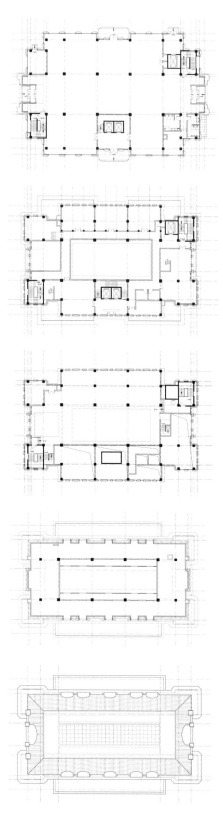

right architectural detail/建筑细部
far right club photo/会所实景照

left　club photos/会所实景照
top right　local club/会所局部
bottom right　club elevations/会所立面

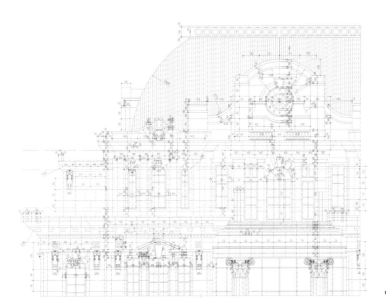

BEIJING MIYUN FLOWER CITY TEMPORARY SALES BUILDING

北京密云国际花都临时售楼处

用地面积	建筑面积	设计时间	建筑类型
7.56万平方米	20.87万平方米	2010年8月至今	商业及中高层住宅

该项目位于密云区城后街与新西路之畔，隶属于密云西区。作为临时售楼处，必须满足快速建造、经济合理、高品质感等多种要求。建筑巧妙运用双层表皮的概念，内层表皮是古典样式的外墙，但采用最为经济的涂料，外层表皮采用花格栅、磨砂玻璃和局部石材等古典建筑元素，两层表皮有机地融合在一起，古典与现代、细腻与精致，优雅地展现在人们面前。特别是在晚上，内层泛光照明进一步凸显两层表皮的层次感和空气感，产生一种朦胧的意境。空间、细节、表皮赋予建筑的不只是对建筑经典设计的追忆，并赋予它在不同光线的交互作用下给人一种光影流转的视觉享受，带给建筑一种特别的动态品质。

The project is located by the side of Chenghou Street and Xinxi Road in Miyun District, belonging to Miyun West Zone. As a temporary sales building, it must possess several characteristics, such as fast construction, economic feasibility and high quality, etc. The building skillfully adopts double-skin concept, where the inner layer is a classical external wall but is painted with economic coating, while the outer layer is made of patterned grid, ground glass and stone at local position to show the elements of classical building. Both layers are organically organized to integrate classical and modern characteristics, delicacy and refinement. Especially at night, flood lighting designed on the inner layer enhances the visual impact of double-layer skin and ventilation effect, producing a filmy feeling. Space, detail and skin not only evoke classic architectural design, but also provide a visual experience in light and shadow generated by interaction among different lights, realizing a special dynamic effect.

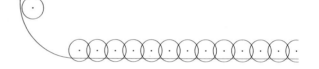
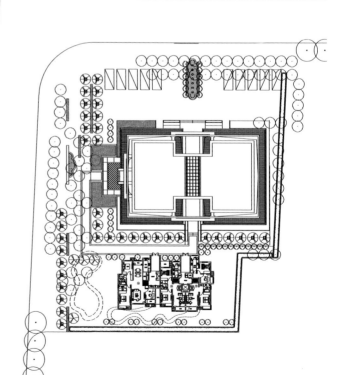

left　site plans/总平面图
below　club facade/会所立面

left   local club facade/会所立面局部
above   local club/会所局部
below   corner of the indoor/室内一角

JINGHU NEW CITY EXPERIMENTAL SCHOOL

镜湖新城实验学校

用地面积	建筑面积	设计时间	建筑类型
32 402平方米	22 020平方米	2009年	文化教育

镜湖新城实验学校位于芜湖市镜湖世纪城5号地块，为周边住区提供了优质的教育配套。学校入口由实验楼、大礼堂和风雨操场半围合而成了生活广场，营造了师生课余时间休憩交流的场所，形成了每个孩子在这里共同生活成长的记忆。学校东侧的静区是由教学部围合的院落，它的灵感来自于东方的书院，庭院内枝繁叶茂的智慧之树下，孩子围坐着听老师传道授业解惑。教学楼的外廊运用了柠檬黄、青草绿作为主色调，每层变化的出挑阳台不仅活跃了形体造型，更让孩子们的课间生活有了趣味。实验楼、风雨操场、行政楼和教学楼出自一种设计手法，彼此间又有气质的不同。简单纯净、活泼有趣，就是镜湖新城实验学校设计的初衷。

Jinghu New City Experimental School is located on Jinghu Century City Plot 5 in Wuhu City and provides high-quality educational supporting facilities for surrounding areas. At the school entrance, laboratory building, auditorium and outdoor playground surround a life square, providing a public space for teachers and students to communicate and rest after class and for children to enjoy themselves. The silent zone at east of the school is a courtyard formed by teaching buildings and the design concept is inspired from the oriental Academy. Wisdom trees are planted in the courtyard, providing a peaceful space for students to learn knowledge. External corridor in the teaching building is painted with lemon yellow and grass green as dominant hues. Diversified cantilever balconies on each floor realize vigorous building appearance and provide more interests for children during break time. Laboratory building, outdoor playground, administration building and teaching building are designed with the same concept, but also remain different characteristics. Simplicity, purity, vigor and interest are design concepts of the Jinghu New City Experimental School.

above   brid's eye view/鸟瞰图
right   middle school photo/中学部实景照

above　primary school inner court photo/小学部内院实景照
below　primary school first to thrid floor plans/小学部一至三层平面图
right　primary school photo/小学部实景照

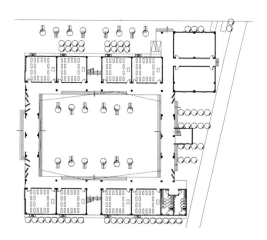 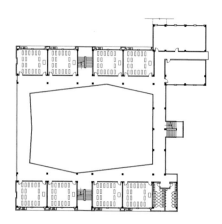 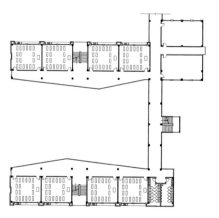

KUNSHAN 21 NEW CITY KINDERGARTEN AND SCHOOL

昆山21新城幼儿园及学校

用地面积	建筑面积	设计时间	建筑类型
48 506平方米	34 214平方米	2009年至今	教育

below kindergarten site plan/幼儿园总平面图
top right bird's eye view/鸟瞰图
bottom right exterior view/外景图

幼儿园的设计以幸运草为灵感，自由生长的四片叶子提供了可灵活分布的四个组团空间。靠近主入口的一片为辅助及办公空间；另三片叶子则以6~9个班为一组围合，拥有各自的专用教室和音体教室，便于教学和活动的分布安排。四片叶子中心是门厅和家长等候区域，主入口处的草茎则是一面嵌有儿童活动主题的弧形剪影墙，将人们从城市道路导入幼儿园门厅。

学校用地的西侧以弧形小区道路为界面。建筑的沿街形态成为设计难点。教室功能的理性要求与弧形边界所形成的张力使五个院落空间有机错位，形成了与基地形状有机契合的空间形式。此外，二层共享平台的引入创造了更多的师生交流机会和可以停留的趣味空间。

The kindergarten is designed according to the shape of four-leaf clover, and the four freely-growing leaves realize flexible distribution of four blocks. The leaf near the main entrance is designed as auxiliary and office spaces, while the other three leaves are designed as classrooms, among which six to nine classrooms form a group to provide dedicated classroom, music classroom and sports classroom, so as to be convenient for arrangement of teaching and sports activities. Hallway and parents waiting area are designed at the center of the four leaves. The caulis at the main entrance is designed into an arch picture wall on which children's activity themes are exhibited, leading people from the urban road to the hallway of kindergarten.

The west boundary of the site is defined by an arch community path. Building appearance along the street is a design difficulty. Rational requirement of teaching classroom and flexibility of arch boundary realize organical stagger layout among the five courtyards, creating a spatial form favorably adapting to the site shape. Furthermore, a shared platform designed on the second floor provides more interesting spaces for teachers and students to communicate and rest.

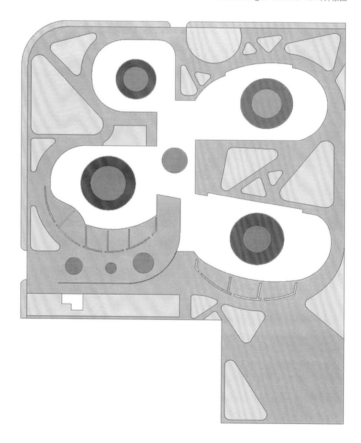

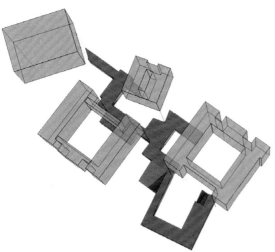

top left   bird's eye view/鸟瞰效果图
bottom left   concept diagram/概念图解
above   exterior view/外景效果图

KUNSHAN XINXIUYI KINDERGARTEN

昆山新绣衣幼儿园

用地面积	建筑面积	设计时间	建筑类型
14 535平方米	15 704 平方米	2011年12月	文化教育建筑

村庄/消融：设计借鉴传统村庄格局，以9个盒子消解30个班级的大体量并形成相对独立的单元式聚落空间。有机的聚落产生趣味场所并形成多层次公共空间，尺度也更适合儿童活动要求。

园林/游廊：9个生活单元和公共空间被一条曲折的"廊子"串联，这条路径来源于对苏州古典园林网师园的解读，悬浮在空中的路径在林霭与光线中穿行，转折处产生微妙的空间感受和对景。

表皮/绣衣："绣"字出现在本案名称以及当地传统苏绣工艺中，由此得到灵感形成建筑表皮，彩色竖向遮阳板如细密的针脚编织成霓裳，包裹整个建筑。简洁而富于色彩变化的表皮使各单元形成差异化的性格特点，契合儿童心理，实现"糖果村庄"的原初设计想法。

Village/Melt

The design makes references to the layout of traditional village, filling 30 classrooms in nine buildings and forming a relatively independent unit-type settlement space. Organic settlement creates interesting locations and diversified public spaces, whose scales are appropriate for children activities.

Garden / Veranda

Nine living units and public space are connected with each other through a piece of zigzag "veranda", which is inspired from the classical garden of Suzhou, the Master-of-Nets Garden. The path floating in the air spreads out in forest fog and sunshine, providing delicate spatial and visual experiences at turnings.

Skin / Embroidery

"Xiu" (embroidery) is used in the name of this project and it is a skill in local traditional embroidery process of Suzhou. The design concept is inspired from embroidery which is used for decoration of building skin. Colorful vertical sunshades are like fine and dense pinpoints which weave a piece of beautiful cloth wrapping the whole building. Simple but colorful appearance forms different characteristics of each unit, satisfying children's mind and realizing original and preliminary design concept of "candy village".

布局
· 家族聚落
· 有机格局

文脉
· 苏州园林
· 地域性

表皮
· 苏绣
· 绣衣镇

left comparative analysis with the Master-of-Nets Garden /与网师园的对比分析
right structure and bird's eye view/鸟瞰图和组织结构

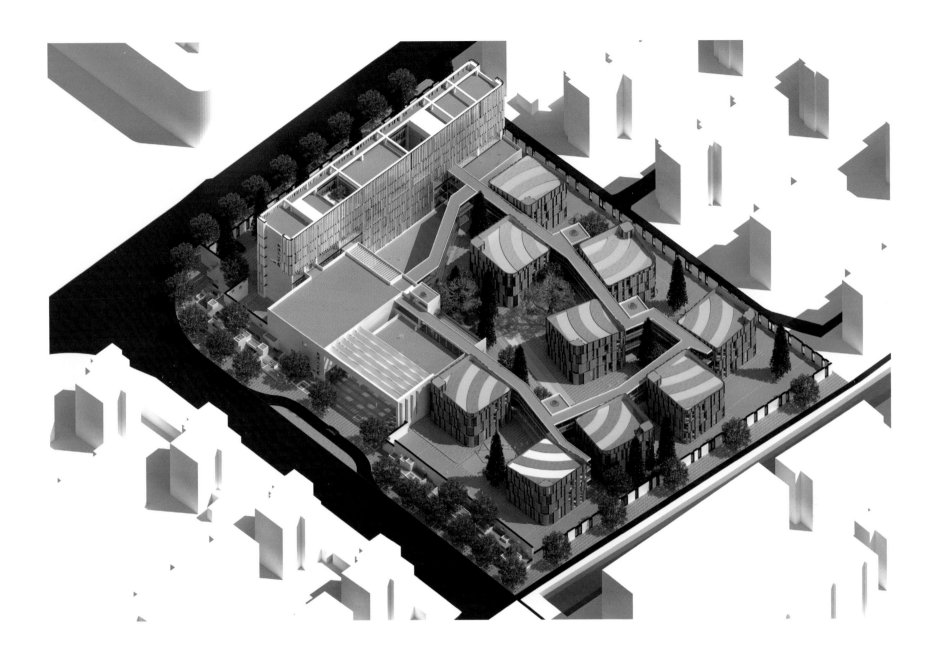

corridor/走廊 + class room units/班级单元 + entrance hall/入口大厅 + administrative office building/行政办公楼 + fencing/围墙

top left   bird's eye view/鸟瞰图
top right   perspective/透视图
bottom left north facde/北立面
bottom right west facde/西立面

i-TOWN RECEPTION CENTER

绿地•未来城接待中心

用地面积	建筑面积	设计时间	建筑类型
1538平方米	2471平方米	2012年3月	售楼处

绿地•未来城接待中心位于南昌高新区瑶湖东岸，环境优美。建筑运用三维数控铸模的技术及GRFC等新材质，以"动态构成"的设计语言，阐述了流动自由、非标准、不规则等设计理念，体现了动态建筑与场所的共生、对话。

由于功能的动态分布，内部空间模式呈现出一种流动的感受，每个空间都有自身的特点，每一个面都可能反映不同的视觉特质，内部空间形态呈现出多样化和个性化特征。建筑空间以一种拓扑的空间姿态展现，空间形态不再均质，空间话语流露多元、复杂化，时间因素在这里得到空间回应，传达出一种动态的视觉张力和流动的空间感受。

白色弧形流动的外观，模糊了面与面之间的分隔界限，使之融为一体。在这仿佛时空交错的空间，观者视点的移动随着动感的表皮流线更显流畅，多画面的构成让人产生一种连续、非固化的视觉享受。

建筑如同一个巨大尺度的未来飞行物停泊于城市之中，其艺术性、科技感的增强，景观效应等等随之扩展，形成了强烈的"向心聚合"效应。

Greenland Future City Reception Center is located at the east bank of Yaohu Lake in Hi-Tech Zone of Nanchang and enjoys beautiful natural environment. The building adopts 3D numerically-controlled molding technique, GRFC material and "dynamic structure" design language to show the design concept of free flowing, nonstandard and irregular shape and to emphasize the coexistence and dialogue between dynamic building and location.

Due to the dynamic function layout, internal spaces give people a flowing feeling with each space having its own features. Each face could reflect different visual effects to realize diversified and personalized internal spatial forms. Architectural spaces are organized in a topological layout to produce non-homogeneous spatial forms, diversified and complicated spatial dialogue. Time factor is reflected in spatial design to form a dynamic visual effect and flowing spatial feeling.

White arched flowing shape blurs the interface among different sides and integrates all sides into a whole. In this special space where space and time are interlaced, moving viewers will see more smooth appearance and integration of multiple images will produce a continuous and non-solidified visual effect.

The building is like a huge future UFO landing in the urban area. Its outstanding artistic and technological characteristics and solemn landscape form strong "centripetal and polymeric" effects.

below   plans/各层平面图
top right   exterior photos/外部实景照
bottom right   interior spaces photos/内部空间实景照

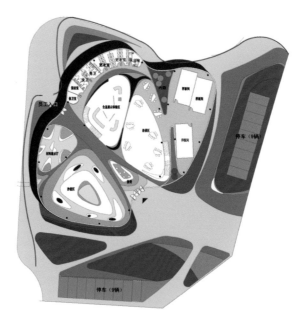

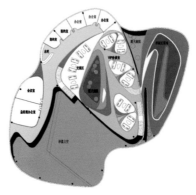

above and right exterior photos/外部实景照
left interior spaces photos/内部空间实景照

4 居住建筑
Residential Building

住宅是最贴近人生活的空间，因此住宅建筑的设计仅满足人的基本需求不难，但要上升到满足人的精神需求层次则不易。人的社会属性决定了社区规划必须从行为模式出发，在交通、景观、配套等层面满足居民的需求，并使居民产生安全感和归属感。人的文化属性要求设计契合地域传统居住模式，在空间结构、空间尺度、建筑体量、立面材质等层面使人产生亲切感和舒适感。作为全国住宅领域的领军设计力量，UA致力于在中国这个世界最大的住宅设计市场中设计出既符合传统居住模式又满足现代生活需求的中国现代居住建筑，努力创造出具有场所精神的居住空间。另一方面，投资商与使用者双方利益的平衡也是UA所关注的。

Residential building is a space nearest to human's life. Therefore, design of residential building can easily satisfy basic requirements of human being, but it is difficult to satisfy people's spiritual requirements. Social attribute of human being determines that community planning must consider people's behavior mode and satisfy residents' requirements for traffic, landscape and ancillary facilities, so as to give residents the sense of safety and belonging. Cultural attribute of human being requires that architectural design shall comply with local traditional residential mode, giving residents intimate and comfortable feelings in the aspects of spatial structure and scale, building volume and facade material. As a leading force in domestic residential building design field, UA dedicates itself to the design of modern residential building which complies with traditional residential mode and satisfies the requirements of modern life in China which has the largest residential building design market in the world, trying to create residential space with location spirit. Furthermore, UA also pays attention to the interest of both investors and users.

SHANGHAI POLY CULTURE FOREST

上海保利·叶上海

用地面积	建筑面积	设计时间	建筑类型
40.83万平方米	71.97万平方米	2007至2010年	住宅

该项目位于上海市宝山区顾村镇，南向紧邻上海最大的城市公园顾村公园，是上海市2007年土地市场总价最高的一宗土地，社会影响较大。规划设计是UA"开放社区、封闭组团"、"三级物管"等规划理念的典型代表。城市道路和内部开放道路构成了整个社区的骨架，将整个地块分隔成五个既相对独立又相互关联的高层组团和三个低层组团。每个组团实现封闭管理，内部完全无车行，又同时共享整个大社区的氛围和配套服务设施，真正实现"住宅单元"—"可防卫组团"—"大社区"三个级别的物管体系。南侧和西侧三个地块布置类独栋别墅，北侧和东侧地块布置独创型高层住宅，整体形成南低北高的优美的空间形态。低层住宅采用两种创新联排户型，打破传统联排住宅行列式的布局形式，使得每户均可享受至少三个采光和景观面，真正实现联排独栋化的设想。设计的创新得到了市场的检验，本项目2009、2010年连续两年成为上海单盘销售冠军，并引领出上海房地产市场追求产品创新的风潮。

Shanghai Poly Culture Forest, located in Gucun Town of Baoshan District of Shanghai City, is adjacent to Gucun Park at south, which is the largest urban park in Shanghai. Total market price of this project site is the highest one in Shanghai in 2007, having extensive social influence. The planning design is a typical demonstration of UA's planning concepts, such as "open community and closed block" and "three-grade real estate management", etc. Urban road and internal open path form the general layout of this community, dividing the whole site into five high-rise blocks which are relatively independent from each other but mutually connected and three low-rise blocks. Closed-off management is adopted for each block to realize vehicle-free internal area and share the community atmosphere and auxiliary facilties. Three-grade real estate management system is successfully realized in "residential unit", "defensive block" and "big community". Detached villa is arranged on the plots at south and west, unique high-rise residential buildings are designed at north and east, forming a graceful spatial layout higher at north and lower at south. Low-rise residential buildings are designed in two innovative townhouse units, breaking the traditional residential townhouse layout in rows. In this way, each unit could have at least three lighting and landscape sides, realizing the concept of detached townhouse. Innovative design has been accepted by the market. This project became the best-selling individual project in Shanghai for two consecutive years in 2009 and 2010, leading the tide of product innovation in real estate market in Shanghai.

below　site plan/总平面图
right　high-rise building exterior view/高层外景图

left and right　high-rise building exterior view/高层外景图

below   villa elevation/别墅立面图
bottom club exterior view/会所外景图
right   villa exterior view/别墅外景图

SHANGHAI POLY LEAVES WHISPER

上海保利·叶语

用地面积	建筑面积	设计时间	建筑类型
24.17万平方米	46.99万平方米	2009至2011年	住宅

该项目位于上海市宝山区顾村镇，顾村公园北侧，紧邻M7号轨道交通，是保利·叶上海的姊妹楼盘，二者共同提升了整个顾村公园板块的城市风貌和住区感受。在整个地块北侧和东侧布置高层住宅，南侧布置创新类独栋联排别墅，整体呈环抱式布局。高层区和别墅区形成南向全景观、无遮挡的构图关系，同时也构成了清晰而有力的城市界面。低层住宅提供两种创新联排户型，在保证高容积率的前提下，创造出完全类独栋的空间效果，取得了良好的市场效果，成为上海市2011年度单盘销售冠军。

Shanghai Poly Leaves Whisper, located at north of Gucun Park in Gucun Town of Baoshan District in Shanghai City, is adjacent to rail transit line M7 and is a companion project of Shanghai Poly Culture Forest. These two projects jointly improve the urban landscape and residential experiences in the whole Gucun Park area. High-rise residential buildings are designed at north and east of the site and innovative detached townhouse villas are designed at south, forming circular layout. High-rise area and villa area face to the panoramic landscape at south, clearly defining the urban boundary. Low-rise residential buildings provide two innovative townhouse units and successfully create a spatial experience which is completely similar to detached villa. Favorable market effects are obtained and this project also became the best-selling individual project of Shanghai in 2011.

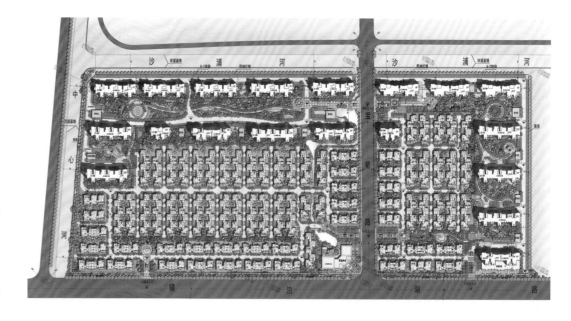

above site plan/总平面图
right villa exterior view/别墅外景图

top left　villa photo/别墅实景照
bottom left　villa night view/别墅夜景图

182

top right   villa photo/别墅实景照
bottom right   villa night view/别墅夜景照
right   villa elevations and photos/别墅立面和实景照

left and above villa exterior view/别墅外景图
below roof detail/屋顶细节
right north elevation/北立面图

SHANGHAI NEW JIANGQIAO CITY

上海新江桥城

用地面积	建筑面积	设计时间	建筑类型
33.21万平方米	60.25万平方米	2010年9月至今	高层住宅

新江桥城作为上海保障性住房八大基地之一，其设计的初衷在于为市区迁居于此的居民提供令他们有归属感的、能安居乐业的宜居社区。

还原"老上海"的居住模式成为规划的出发点。"街—弄—院—宅"这一传统守望相助的邻里模式启发了创新四级物管的"可防卫邻里"组团规模的形成。便捷的仿生蛛网结构基于对微距服务的老上海弄堂口烟杂店的再现，并有效串接起各组团，将商业、中小学、幼儿园、社服中心等配套设施按合理服务半径科学布点，服务居民。将慢行交通、弹性停车、可接触景观等先进规划理念引入社区，提供了超越老上海的现代舒适人居环境；有价值感和尊严感的建筑立面和全功能配置的合理户型使居民对社区产生了深深的归属感。

New Jiangqiao City project as one of the eight large bases of indemnificatory housing in Shanghai is designed with the purpose to provide a sense of belonging and a residential community in peace and contentment to residents moving here from downtown area.

The design concept is to reproduce the residential mode of "Ancient Shanghai". The innovative four-grade real estate management mode of "defensive neighbourhood" block scale is inspired from traditional residential mode of "street—lane—courtyard—house" in Shanghai. Convenient bionic web structure of this project reproduces the cigarette store providing micro-distance services at lane mouth in old Shanghai and effectively links all blocks. Supporting facilities, such as commercial building, middle school, elementary school, kindergarten and social service center, etc., are scientifically arranged in a reasonable service radius to serve local citizen. Advanced planning concepts, including slow traffic, elastic parking and touchable landscape, are introduced into community planning to provide modern and comfortable habitat environment better than the old Shanghai; building facades show solemnity and elegancy and reasonable unit layout is configured with full functions, producing strong sense of belonging to citizen.

below site plan/总平面图
right bird's eye view/鸟瞰图

left and right exterior view/外景图

left and right exterior view/外景图

SHANGHAI POLY PHOENIX TREES WHISPER

上海保利·梧桐语

用地面积	建筑面积	设计时间	建筑类型
6.763 5万平方米	8.027 8万平方米	2009年12月	高层住宅与别墅

below bird's eye view/鸟瞰图
right detached villa photo/类独栋别墅实景照

该项目是保利集团开发的重要项目之一，位于上海市嘉定区和政路以东，嘉戬公路以北，周边公共设施齐全，交通便利。东西走向的蜡烛河穿越其中，将地块分为南北两个地块，北部环境清幽，全部为该地块定制的类独栋别墅产品，南部以高层住宅为主，整体建筑采用英式官邸风格，大气典雅。

Shanghai Poly Phoenix Trees Whisper project is one of the important real estate projects developed by Poly Group. The project is located to the east of Hezheng Road and north of Jiajian Highway in Jiading District of Shanghai City, enjoying completed public facilities and convenient traffic systems. Candle River flowing along west and east passes through this project site and divides the site into north and south parts, among which the north part enjoys silent and peaceful environment and is designed with detached villas which are custom-tailored for this area; while the south part is configured with high-rise residential buildings of British elegant mansion style.

below and right   detached villa photos/类独栋别墅实景照

XI'AN CHANBA INTERNATIONAL ECHO-CITY

西安浐灞国际生态城

用地面积	建筑面积	设计时间	建筑类型
323 824平方米	319 383平方米	2009年至2010年5月	住宅

西安浐灞国际生态城位于西安市主城区东南部的，浐灞生态区南端。本项目作为国际化游憩生态新城的一期工程，以自然、现代、生态为理念，打造生态宜居社区。规划上充分保留和利用浐河水系打造滨水生态轴，并从滨水生态轴引入一条景观轴线连接东西地块。设计结构本着最大程度地利用景观资源，结合自然地形的原则，地块由东向西分别布置纯步行的全景联排别墅以及全景高层住宅组团。通过现代的三级物管模式和微距式社区服务体系，提升了生态城的社区归属感。

Xi'an Chanba International Echo-City is located at south of Chanba Ecological Zone at southeast of downtown area of Xi'an City. This project as phase I of a new international ecological city is designed according to natural, modern and ecological concepts, with the purpose to create an ecological and habitable community. The planning reserves and utilizes the Chanhe River to create a waterfront ecological axis, from which a landscape axis is designed to link the east and west plots. The principle of structural design is to realize maximum utilization of landscape resources with consideration of natural terrain. Panoramic townhouse villas and panoramic high-rise residential blocks are arranged from east to west on the site, where only pedestrian paths are designed. Modern three-grade real estate management mode and micro-distance community service system are adopted to enhance community attachment sense of the ecological city.

above   bird's eye view/鸟瞰图
right   architectural photo/建筑实景照

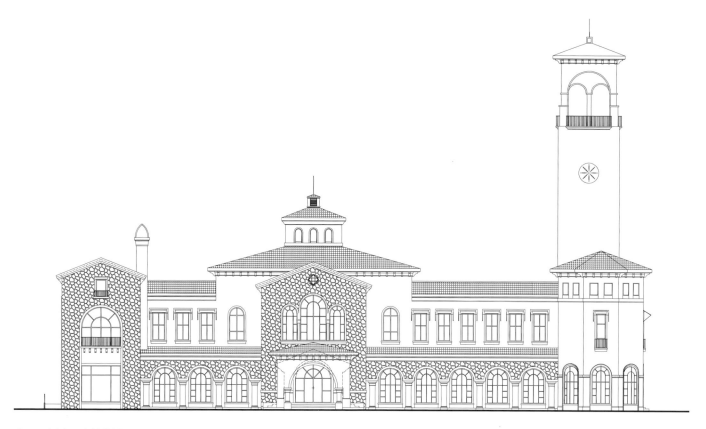

above club facade/会所立面
below club photo/会所实景照
right architectural photo/建筑实景照

POLY YE ZHILIN

保利叶之林

用地面积	建筑面积	设计时间	建筑类型
98 805.1平方米	266 687.95平方米	2011年2月至今	住宅

本案位于上海市宝山区，是保利集团上海公司2011年重点打造的高品质楼盘。地块拥有优越的公园景观和河道景观。规划结构遵循"使小区的每一户人家都享有丰富的景观资源"的原则，通过内部东西向的主景观绿轴将三块地紧密相连，形成独有的网格化景观体系。建筑与景观和谐相融，成为闹市中难得的净土。

在产品设计中，对经典"90"户型进一步优化，增加了厨房和卫生间的实用性，为住户的生活提供既紧凑又舒适的空间。在造型设计中，将西式建筑的比例关系与东方建筑的细部元素相结合，创造出简洁大气，且渗透着东方韵味的新颖的住宅建筑风格。

This project is located in Baoshan District of Shanghai City and it is a high-quality real estate project created by Poly Shanghai Real Estate Development Co., Ltd. in 2011. The project site enjoys advantageous park landscape and watercourse landscape. The planning structure follows the principle of "providing abundant landscape resources for every room in the community". Three blocks are closely connected together through an internal main landscape green axis spreading out to east and west. Harmonious coexistence is realized between buildings and landscape, producing a rare pure land in busy downtown area.

During product design, the designers further optimize the classic "90" room layout and add the practicability of kitchen and washing room, creating compact and comfortable spaces for users' daily life. On the aspect of appearance design, the designers combine the proportional relation of western building with detail elements of oriental building together to realize a simple, generous and novel residential building style, showing oriental beauty.

above site plan/总平面图
right residential perspective/住宅透视图

left residential perspective/住宅透视图
right architectural detail/建筑细部

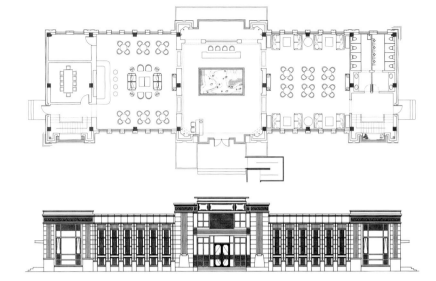

left   club perspective/会所透视图
above club plan and facade/会所平面和立面图
below club perspective/会所透视图

LINGANG NEW CITYSOCIAL HOUSE

上海临港新城限价房

用地面积	建筑面积	设计时间	建筑类型
60 984平方米	104 638平方米	2011年8月	办公、会展、会议

　　本案是位于临港新城这一以国际先进理念打造的现代化卫星城中的限价房项目，旨在探索新的社会历史背景下保障房设计的新可能。

　　根据控规的路网尺度和城市设计的形态要求，突破了我国当前常见居住社区由数幢独立建筑围合成为组团或小区的空间模式，以连续不断的建筑沿街界面完全围合而再现了久违的真正街道空间，使街道回复了昔日作为人活动场所的尺度并实现了令人流连忘返的街道美学。在满足全明户型、厨卫相邻且每个开间至少2.2米以上、立面消灭深凹槽的小户型要求基础上，本案创造性地实现了更适宜人生活的、充满活力和场所感的现代社区新模式。

Shanghai Lingang New City Social House project is located in Shanghai Lingang New City which is a modern satellite town developed with internationally advanced concept, and purpose of this project is to explore new possibility of social house design under new social and historical backgrounds.

Based on road net scale of controlled planning and requirements of urban layout, designers successfully break through the spatial form of building cluster or residential community enclosed by several detached buildings which are usually adopted in current residential communities, and reproduce a true and completely closed street space by arranging consecutive buildings along street. Streets become public spaces for people's activities and realize street aesthetics, giving deep impression to people. This project adopts small room layout. All rooms which are designed with large windows, kitchen and washing room are adjacent to each other, each opening is at least 2.2m, and deep grooves are set on facades. Furthermore, this project creatively realizes a new modern community mode with better habitability, more vigor and larger space.

above　site plan/总平面图
right　bird's eye view/鸟瞰效果图

 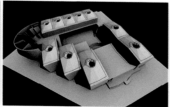 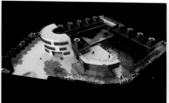  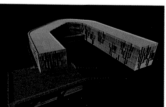

above perspecitive/透视图
below bird's eye view and contest project/鸟瞰效果图和比选方案

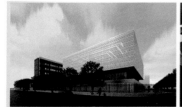    

SHANGHAI XINLI WILLIAN MANSION

上海新里威廉公馆

用地面积	建筑面积	设计时间	建筑类型
79 165.7平方米	193 448平方米	2011年1月至今	住宅

项目用地位于上海市嘉定区，毗邻嘉定文化古迹"古漪园"。整体规划以"人性、人文、宜居"为指导理念。尊重城市历史与文脉，尊重基地周边原有的自然人文价值，强调合理的人性空间、居民归属感，强调空间的整体匹配性。基地位于居住区与具有文化价值坐标的"古漪园"之间，遵循城市辐射理论，以柔性的方式融入相对应的城市密度辐射圈中，强调对城市现有节点的呼应以及对文化历史现状的尊重。

Shanghai Xinli Willian Mansion project is located in Jiading District of Shanghai City and is adjacent to the "Guyi Garden", a historical cultural relics in Jiading District. The general planning is based on the concept of "humanity, culture and habitability". This project shows respects to urban history and culture, existing natural and cultural values around the site, and makes emphasis on reasonable humanistic space, sense of belonging and overall spatial harmony. The project site is between a residential community and the "Guyi Garden" which is a cultural landmark. Following the urban radiation theory, this project is integrated into related urban density radiation circle through a flexible method, so as to emphasize response to existing urban nodes and show respect to existing cultural and historical elements.

left site plan/总平面图
right bird's eye view/鸟瞰图

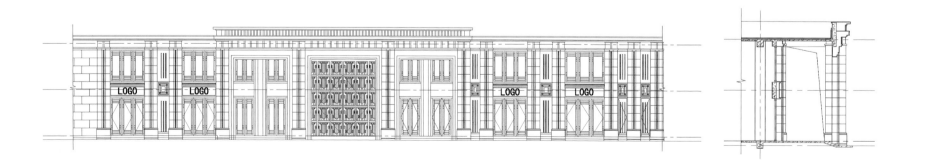

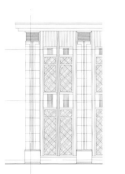   

top landscape of central axis—clubhouse & entrance of main hall view/
中央景观轴—会所及大堂入口效果图
below clubhouse & entrance of main hall facade & wall node/
会所及大堂入口立面及墙身节点

above entrance of main hall view/入口大堂场景图
below villa perspective /洋房透视图
right high-rise residential building perspective /高层住宅透视图

GREENLAND YISHU

上海绿地壹墅

用地面积	建筑面积	设计时间	建筑类型
131,487.7平方米	262,479.9平方米	2010年10月至今	混合开发

本案在设计中引入"深蓝格"的概念。"深蓝"意为"水之深蓝",强化水在整体规划布局中的统领性涵义。中央大水景周边布置船坞式小型商业单体;强调本案所在地朱家角水乡的地域性特征,打造具有水乡气质的特色商业。"格"——规划结构上采用网格扭转方式,在基地中找到一种清晰的逻辑关系,这种逻辑高效灵活。

Shanghai Greenland Yishu Villa project introduces the "deep blue grid" concept into the architectural design. "Deep blue" means "the navy blue color of water" and emphasizes the guiding role of water in general planning layout. Dock-like small individual stores are arranged around the large central waterscape, showing the regional feature of the waterside Zhujiajiao Town where this project is located. The design concept is to create a waterside featured commercial project. "Grid" means that planning structure adopts twisted grid to generate a clear logical relation, which is highly effective and flexible.

1. continuous enclosed space, courtyard in different directions, central buildings distribute
2. enclosed courtyard like windmill, the central space is more open
3. enclosed courtyard interconnected, open to the central main loop
4. courtyard open to the central main loop, central buildings are grouped, sight line is more open

1. 连续的围合空间,院落方向不同,中央建筑散布。
2. 风车形围合院落,中央空间更开敞。
3. 围合院落相互联系,向中央主环路开放。
4. 院落向主环路开放,中央建筑成组,视线更通透。

left site plan/总平面图
below bird's eye view/鸟瞰图

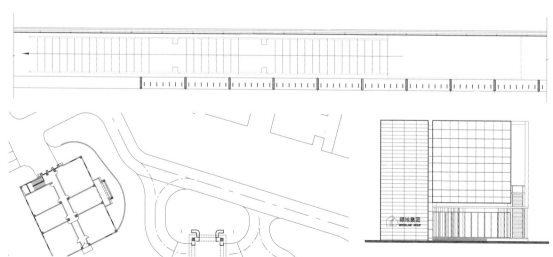

above sales building final form/售楼处最后造型
left detail、facade/详图、立面图
right detached commercial building entrance/独栋商业入口场景

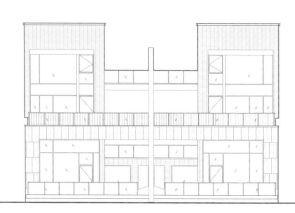

top left   detached commercial building perspective /独栋商业效果图
bottom left   detail、facade/详图，立面图
below   single rendering /单体透视图

BEIJING SISLEY CHATEAU

北京西斯莱公馆

用地面积	建筑面积	设计时间	建筑类型
163 983平方米	401 500.7万平方米	2009年7月至2010年12月	住宅

　　北京西斯莱公馆位于北京市大兴区，东至辅高路、南至金星路、西至兴丰大街、北至后高路。地块周边有大型商业及医院，辅以城市公共轨道交通，生活配套设施齐备，适宜生活居住。

　　规划采用典型的"开放社区、封闭组团"、"三级物管"等规划理念，通过一条开放的林荫大道将不同居住产品的住宅组团有机串联在一起。形成各自独立又相互共享大社区配套设施和氛围的六个组团。建筑设计尝试住宅外观公建化，体量上遵从古典横竖三段式的构图法则，外立面全部采用封闭阳台的做法，营造出稳重内敛、雄浑大气的不同以往的住宅形象。项目建成后基本达到设计之初预想的两个目的：大大提升了大兴新城区的城市面貌和居住体验；作为上海绿地集团进入北京的第一个项目，实现了品质和销售的双重成功，成为北京市2010年的单盘销售冠军。

Beijing Sisley Chateau project, located in Daxing District of Beijing City, is adjacent to the Fugao Road at east, Jinxing Road at south, Xingfeng Avenue at west and Hougao Road at north. There are large commercial buildings and hospitals around the site. Convenient urban public rail traffic and supporting facilities add more habitable elements to this residential project.

The planning adopts typical planning concepts, such as "open community and closed block" and "three-grade real estate management", etc. Six residential blocks which are relatively independent and share supporting facilities and landscape are connected into an organic space through an open boulevard. The architectural design explores residential building appearance in public building. Building volume is designed according to the classical three-segment structure stretching transversely and longitudinally. External facades are designed with closed balcony to create an introvert and solemn appearance, showing an outstanding generous and elegant residential building image. After completion, this project will basically reach the two expected design aims, one of which is to obviously improve urban environment and habitable experience in Daxing new urban area and the other is to realize high quality and best selling as Shanghai Greenland Group's first real estate project in Beijing. The project became the best-selling individual project of Beijing in 2010.

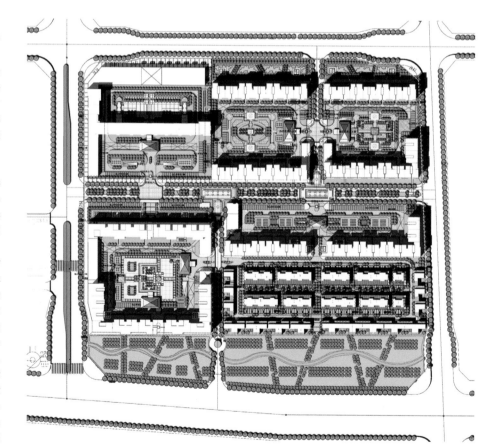

above　site plan/总平面图
right　residential photo/住宅实景照

left and right   residential photos/住宅实景照

BEIJING XINGCHANG YISHU VILLA

北京兴创屹墅

用地面积	建筑面积	设计时间	建筑类型
152 439.2平方米	210 573.2平方米	2010年月至2010年12月	住宅

该项目是由北京兴创公司开发的高端居住社区，地块位于北京市大兴区双高路南侧，交通便利，周边环境幽雅，整个区域比较适合居住。整个地块分为三片区域，东南角为联排别墅区，东侧中间为配套高端会所，其他区域为多层和小高层。立面造型体块清晰，古典大气。建筑细部运用古典元素，彰显建筑的精致与高贵，强调手工感和厚重感，代表了当前法式豪宅的建筑风格。

Beijing Xingchuang Yishu Villa project is a high-end residential community developed by Beijing Xingchuang Investment Co., Ltd. The project site is located to south of Shuanggao Road in Daxing District of Beijing City, enjoying convenient traffic system and beautiful environment. It is a habitable residential project. The whole site is divided into three blocks, including townhouse villa block at southeast corner, supporting high-end chamber block at the middle of east part and multilayer and high-rise block at other areas. Building facades show clear and classical generosity and solemnity. Classical elements are used for detail design to show the delicacy and elegancy of buildings. The senses of handicraft and stability represent the architectural style of French luxurious residential building.

above site plan/总平面图
right district interior court perspective/小区内院透视图

above club perspective/会所透视图

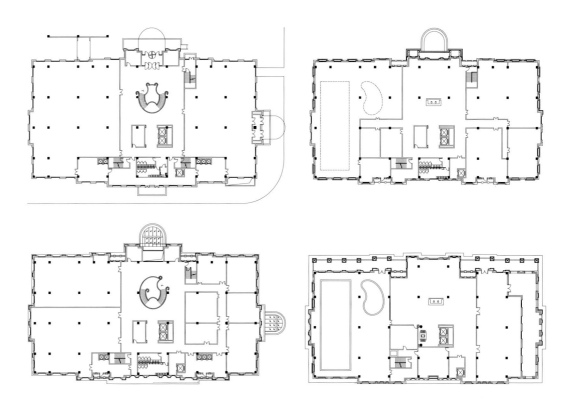

left group interior perspective/组团内景透视图
above architectural level/建筑平面图
below architectural facade/建筑立面图

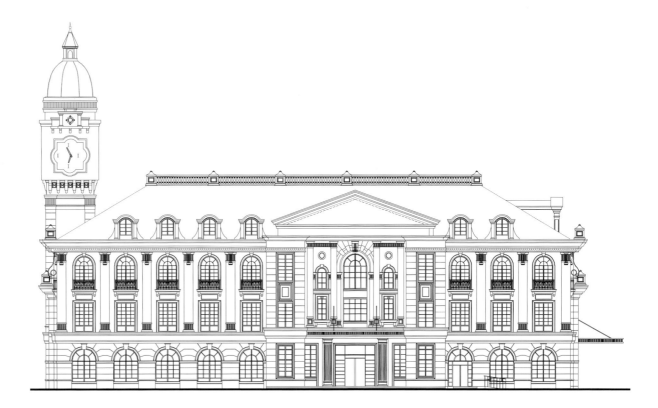

TIANJIN JIXIAN VILLAGE OF ART

天津蓟县星梦工坊

用地面积	建筑面积	设计时间	建筑类型
428 045平方米	178 896平方米	2008年	住宅

天津市盘龙谷文化城位于天津市5A级国家名胜风景区—蓟县西北部的许家台乡，星梦工坊是文化城中的一期项目，该项目西距北京53千米，南距天津110千米。区内地势起伏，多为东西走向的山体，且现状并不完整。

有中国哲学家认为："体现传统的价值的东西是具有超时代的意义，具有高度的稳定性而非随时变易的东西。过去的价值和思想，在这个意义上，不仅属于过去，也属于现在和未来。"传统聚落、民居反映的天人合一、家族至上等价值观，能够弥补现代都市生活所丧失的个人生活的私密性和多样性。传统居住文化中的聚落、庭院、意境等要素所体现出来的多重意义特征适应现代生活精神上的需求。传统文化与生态环境的各种问题时常成为设计的焦点。

本案例有着其特定的自然生态环境、特殊的使用对象以及特定任务，设计观念表达对有关问题的一些思考与尝试。

Tianjin Dragon Valley Cultural City is located in a AAAAA state scenery resort in Xujiatai Town at northwest of Jixian County of Tianjin City. Village of Art is phase I of the project and it is 53km to Beijing at west and 110km to Tianjin at south. The mountainous site terrain is undulate and mainly spreads out from east to west, but its status quo is not completed.

Chinese philosophers believe that "It is ultramodern significance and high stability but not inconstant elements that show traditional values. Ancient values and thoughts, in this aspect, belong to not only the past, but also modern time and the future. Traditional residential values, such as integration between nature and human being, and supreme family attitude, etc., could make up the privacy and diversity of private life which have been lost in modern urban life. Settlement, courtyard and conception elements in traditional residential culture demonstrate multiple meanings and features that could meet spiritual demand of modern life. Various conflicts between traditional culture and ecological environment usually become the key points of planning design.

This project has specific natural and ecological environment, special users and certain tasks, so the design concept gives some thoughts and attempts to solve relevant problems.

left site plan and concept diagram/总平面图和概念图解
right architectural photos/建筑实景照

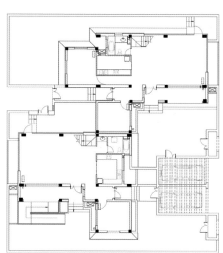

top left and top right　architectural photos/建筑实景照
bottom left and right　CAD plan/CAD 平面图

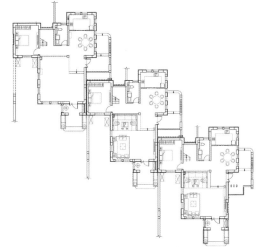

top left and top right   architectural photos/建筑实景照
bottom left and right   CAD plan/CAD 平面图

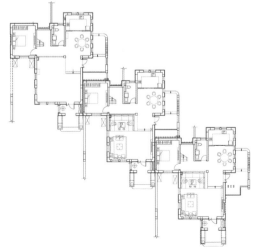

left and right architectural photos/建筑实景照

HUAXIA FORTUNE UNITED KINGDOM PALACE

华夏幸福英国宫

用地面积	建筑面积	设计时间	建筑类型
20万平方米	25.3万平方米	2010年5月	高层住宅+别墅

below site plan/总平面图
right townhouse renderings/联排别墅效果图

该项目由华夏幸福基业开发建设，是主要面向北京市人群的一个大型综合性开发项目，位于河北省廊坊市大厂县，南临150米宽的滨河景观休闲带，与潮白河及天然杨树林隔路相望，更有自然溪流贯穿地块南北。英式的类独栋别墅滨水布置，北部高层则引进次街、三级物管等全新生活理念，领先当地生活模式。东南角的风情商业街是整个项目的点睛之笔。该地块的设计完美诠释了郊区大盘的品位与格调。

United Kingdom Palace project is developed by China Fortune Land Development Co., Ltd. (CFLD) and it is a large complex development project for Beijing market. This project is located in Dachang County of Langfang City in Hebei Province and is adjacent to a 150m-wide riverside landscape leisure belt at south, crossing which it faces to the Chaobai River and natural poplar forest. In addition, a beautiful natural river flows from north to south on the site. British-style detached villa is arranged along riverside and brand-new living concepts, such as secondary street and three-grade real estate management, etc., are introduced in to the high-rise buildings at north to realize a better lifestyle. Flavor commercial street located at southeast corner on the site is a key point of this project. The site planning shows a perfect quality and environment of a large suburb real easte project.

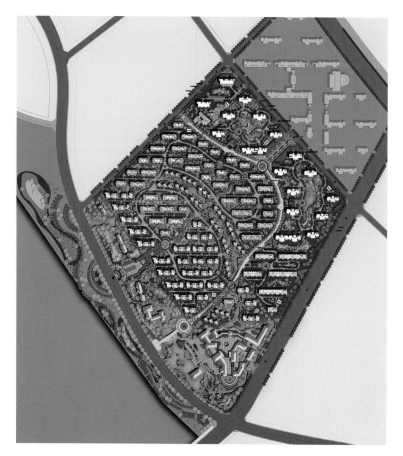

left townhouse renderings/联排别墅效果图
below townhouse elevation/联排别墅立面图

XIANGHE CENTRAL CONSULAT·I HOME

香河中央领仕馆·我家工坊

用地面积	建筑面积	设计时间	建筑类型
约13.84万平方米	14.22万平方米	2010年2月	高层住宅+别墅

项目位于河北省廊坊市香河县，占地约13.84万平方米。本次规划以次街将地块分为南北两部分，北部类独栋别墅产品打造私密的庭院空间，南部的创新小面积高层通过开放道路及封闭组团的设置强调住户情感的培养，营造与众不同的社区氛围。

I Home project is located in Xianghe County of Langfang City, Hebei Province, and covers a total land area of about 138,400m² hectares. The site is divided by a secondary street into north and south parts, the north part is built into a private courtyard space formed by detached villa, while the south part is built with small innovative high-rise residential buildings. Open paths and closed blocks enhance dwellers' emotion and create unique community atmosphere.

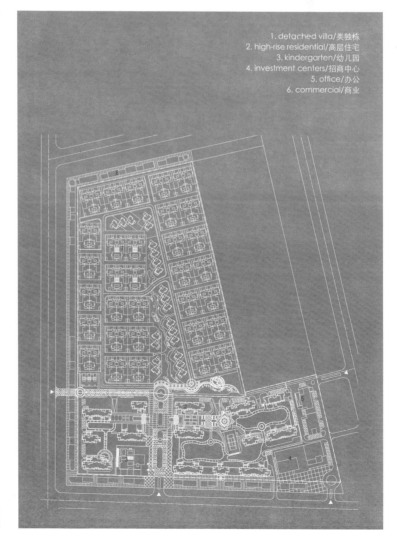

1. detached villa/类独栋
2. high-rise residential/高层住宅
3. kindergarten/幼儿园
4. investment centers/招商中心
5. office/办公
6. commercial/商业

above site plan/总平面图
right high-rise residential photo/高层实景照

left and below   high-rise residential photos/高层实景照

left and right   detached villa photos/类独栋别墅实景照

GUANGZHOU POLY WEST COAST

广州保利西海岸

用地面积	建筑面积	设计时间	建筑类型
91 597平方米	65 539平方米	2009年	别墅

惬意滨水生活：一线珠江，双重水岸，三座公园。最适宜人居朝向：全面考虑岭南居住习惯，围而不封，兼得阳光和清风。这些都是广州保利西海岸的项目标签，作为整体项目的最高端产品，小区采用经典环状道路结构，环路内为双拼别墅，环外为类独栋别墅和联排别墅。立面处理上，借鉴英国都铎式建筑风格，打造纯英式社区，同时为了迎合现代都市大众，将部分烦琐线脚简化，使其既有英国韵味，又不失现代感。

Comfortable riverside life: one piece of Zhujiang River, two riverside spaces and three parks. Best habitable orientation: full considerations are given to the dwelling habit in South China, enclosed but not closed layout provides sufficient sunshine and ventilation. These are features of Guangzhou Poly West Coast project. As the best high-end project of the whole planning, the community adopts typical ring-road system, with two-family-houses inside and detached villas and townhouses outside. Facade treatment makes references to British Tudor architectural style to create a pure British-style community. Meanwhile, to meet the requirements of modern urban citizen, some complicated corners are simplified to reserve British style while realizing modern features.

above site plan/总平面图
right villa photo/别墅实景照

left and right   villa photos/别墅实景照

left　club night photo/会所夜景照
right　club interior photos/会所内部实景照

CHENGDU BAOLY CENTER

成都保利中心

用地面积	建筑面积	设计时间	建筑类型
4.3682万平方米	37.77万平方米	2010年8月	办公及商业综合体

本次规划的基地位于成都市中心黄金地段,美国领事馆以东。地块分为东西两部分。西地块以商业、办公和住宅为主,东地块以商业和Loft为主,容积率达到8.0。设计打造保利在成都市中心最高端的,集超高层住宅与高端商业、办公等多种功能为一体的城市综合体项目。

The planning site is located at a golden location at the urban center of Chengdu city and at the east of American Consulate. The site is divided into each and west parts, among which the west part is mainly designed with commercial, business and residential buildings, while the east part is mainly planned with commercial building and Loft building, reaching a plot ratio as high as 8.0. Design concept of this project is to create the best high-end urban complex integrating super-high residential building, high-end commerce and offices at the urban center of Chengdu.

below　general plan/总平面
right　photo within the street map/内街实景图

1. 超高层住宅/skyscraper
2. 办公/office
3. 商业/business
4. LOFT

left and right photos/建筑实景照

left and top right   photos/实景照
bottom right   inside photo/内部实景照

CHENGDV RVNYANG SUNSHINE GARDEN

成都润扬北城一号

用地面积　　建筑面积　　设计时间　　建筑类型
8.27万平方米　29.6万平方米　2009年　　住宅

润扬北城一号，成都市北部新城唯一的托斯卡纳风情低密度庄园住区，位于市区东北部。采用一环两轴三节点六片区的规划理念，该项目拥有成都首例意式私人会所——康堡：在6 000平方米豪华配置的"康堡"里，不用出社区门，就能和家人、朋友享受与世界同步的时尚，咖啡吧、红酒吧、雪茄吧、健身房、书吧、SPA、壁球馆、室内恒温泳池等铸就品质感的城市庄园生活。

润扬北城一号为"庄园主们"打造3万平方米帕溪水景公园，巧妙结合小区入口外坡形地貌，漫坡缓溪引领回家的路，同时有效将住区和道路完全分离，形成私密、独享的城市庄园生活氛围。润扬北城一号特有的六大主题组团住区，以各式经典意式风情组团形成各不相同的庄园生活体验，在绿野迷踪里和孩子享受天伦之乐，无不体现出一幅真实的庄园生活美景。

Runyang Sunshine Garden, located in the northeast part of Chengdu City, is the unique Tuscany-style low desnity residential manor project in the Northern New City of Chengdu. The planning concept contains one ring, two axes, three nodes and six blocks. Kangbao Garden which is the first Italian-style private chamber is located in this project and covers a luxuriously decorated space of 6,000m². People can share with their family members and friends a high-quality urban manor life supported by internatinally-advanced fashion, such as café, red wine bar, cigar bar, fitness center, bookstore, SPA, squash hall, indoor constant-temperature swimming pool, etc.

Runyang Sunshine Garden creates for users a waterside garden covering an area of 30,000m² along Paxi River. External ramp at the entrance to the community is skillfully treated to create low ramp paths leading home. Meanwhile, dwelling and paths are effectively separated to provide a unique experience of private urban manor life. The specific six theme blocks of this project are desgined in various typical Italian styles to provide diversified manor life experiences. Parents can enjoy a happy time with their children in green environment. People could find beautiful landscape of true manor life everywhere in this project.

below　site plan/总平面图
right　bird's eye view/鸟瞰图

left the main group interior scene/主要组团内实景
right residential scene/住宅实景照

below   photos/实景照
right   entrance bird's eye view/入口鸟瞰图

CHENGDU XINLI PECK RESIDENCE

成都新里·派克公馆

用地面积	建筑面积	设计时间	建筑类型
249 700平方米	约600 000平方米	2005年至2008年	混合开发

成都新里·派克公馆位于成都素有"人文城西，新兴富人区"之称的西高新国际社区，正处在成都高新技术产业开发区西区和金牛区交界的核心地带。项目占地约25万平方米，总建筑面积近60万平方米，其中产品类型有叠拼别墅、电梯洋房、高层住宅等。新里·派克公馆由住宅和公建两部分组成，建筑高度总体呈南高北低以满足航空限高要求。公建设于基地北端，包括商业、餐饮、会所、青年公寓、健身房、游泳馆、篮球馆等，可兼顾社区内外客户服务的需求。住宅部分根据成都当地的气候特点及规划要求，基本上采用围合式的建筑布局，以营造出尺度适宜的景观庭院，增强住户的归属感。建筑造型以影响深远的"装饰艺术"风格为设计灵感。

Chengdu XINLI Peck Residence, located in the West High and New International Community where is praised as "humanistic west urban area is an emerging district for the rich" in Chengdu, enjoys a core location at the boundary between High-Tech Industrial Development Zone and Jinniu District of Chengdu. The project covers an area of about 250,000m², the total floor area being about 600,00m² and consists of townhouse villa, western-style building equipped with elevator and high-rise residential building, etc. Xinli Peck Residence is composed of residential buildings and public structures. General height of buildings is higher at south and lower at north, so as to satisfy the aviation height limitation. Public facilities are built at the north end of the site, consisting of commercial building, restaurant, chamber, youth apartment, fitness center, swimming pool and basketball court, meeting requriements of customer service both inside and outside the community. Considering local climatic conditions and planning requirements, residential buildings are designed in an enclosed layout to court a landscape cortile of appropriate scale and to enhance dwellers' sense of belonging. The design concept of building appearance is inspired from the famous Art Deco style.

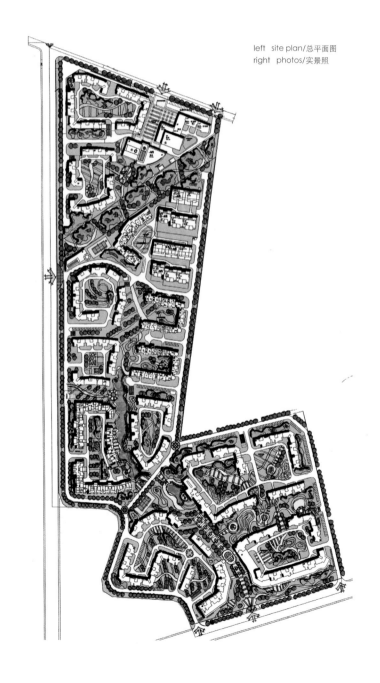

left site plan/总平面图
right photos/实景照

left and right photos/实景照

JIANGYIN AIJIA LA VILLA

江阴爱家·名邸

用地面积	建筑面积	设计时间	建筑类型
11.21万平方米	141.6万平方米	2009年至今	住宅

爱家·名邸位于江阴市敔山湾西部，三面环水，远眺群山，具有优美的自然景观资源。规划通过创新"天"字形结构，将小区分为南北两个组团，强化两组团间的水岸景观，提升小区的居住品质。行为主导交通模式以完全的人车分流方式在提供步行便捷性及安全感的同时在社区内部营造出富有活力的生活空间。三级物管结合星级入口大堂，增进了小区的外在价值感与内在品质感。住宅立面融合现代与古典，在简洁而富于变化的体块关系中通过石材和铝板相结合的精妙细部设计实现了动人的立面效果与现代工艺之美。

Aijia La Villa project, located at west of Yushan Bay of JiangyinCounty, is surrounded by water at three sides and faces to mountains at a far distance, enjoying beautiful natural landscape resources. The planning adopts innovative 天-shape structure ("天" is a Chinese character, meaning the sky) to divide the community into north and south blocks. Waterside landscape between the two blocks is improved and dwelling quality of the community is upgraded. Behavior-guided traffic system separates pedestrain and vehicle flows completely, providing convenience and security for pedestrain and creating vigorous living space in the community. Three-grade real estate management cooperates with star-level entrance lobby to further increase the visual effects and internal functions of the community. Modern and classical elements are integrated into the facades of residential building, and stone and aluminium plates are used for detail decoration on the simple but diversified building volumes, successfully realizing vivid facade effect and aesthetics of modern technology.

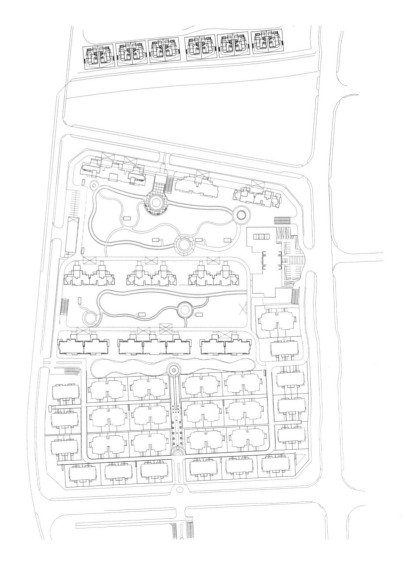

right site plan/总平面图
far right club photo/会所实景照

left　club exterior photo/会所外景实景照
below　local club facade/会所立面局部

left   villa exterior view/别墅外景图
below   elevations and section/立面和剖面图
right   villa exterior view/别墅外景图

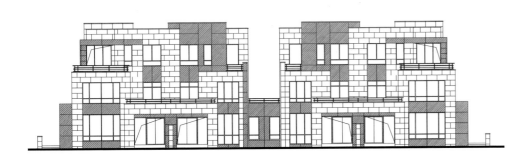

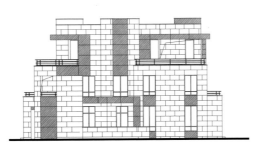

NANJING FORTE EASTERN MANOR

南京复地东郡

用地面积	建筑面积	设计时间	建筑类型
209 420平方米	457 025平方米	2011年5月至今	住宅

本案毗邻紫金山山麓，拥有独一无二的外部自然景观。在规划中，力求在严格的限高及容积率条件下将多层洋房的品质做到极致。设计将古典规划模式"五进空间"运用到整体空间的塑造中，真正做到了建筑与景观融为一体。在产品设计中，创造性地提出了"第五代洋房"的概念，研发了每户电梯独享、每户赠送采光地下空间及庭院分享到户、独立双车位停车库等超附加值的赠送空间。本案在特定规划条件下在新型洋房产品的规划和设计方面做出了创新性的尝试。

Forte Eastern Manor is adjacent to Zijin Mountain and enjoys unique natural landscape. The planning strives to create best western-style building with strict control of building height and plot ratio. The classical planning mode of "Five-Enter Space" is adopted for general space layout and realizes real integration between building and landscape. In design process, the designers creatively put forward the concept of "5th Generation of Western-style House", successfully researched and developed a spatial layout to design one elevator for each suite and provide value-added spaces to each suite, such as lighting basement space, shared courtyard and independent double parking lots, etc. This project makes innovative attempt for planning and design of innovative western-style house under specific planning conditions.

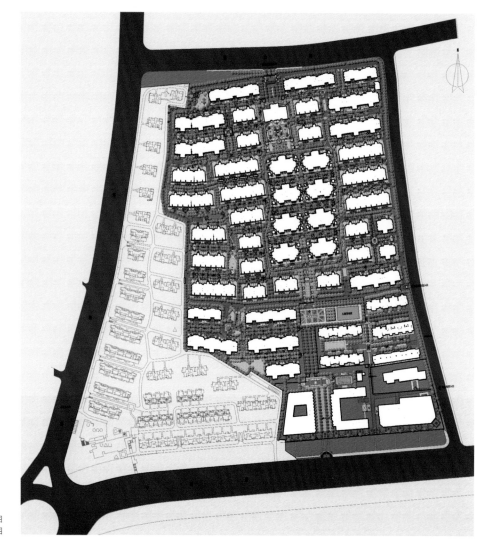

right site plan/总平面图
far right residential perspective/住宅透视图

left and right residential perspective/住宅透视图

below　club elevation and perspective/会所立面和透视图
right　clubhouse entrance/会所入口

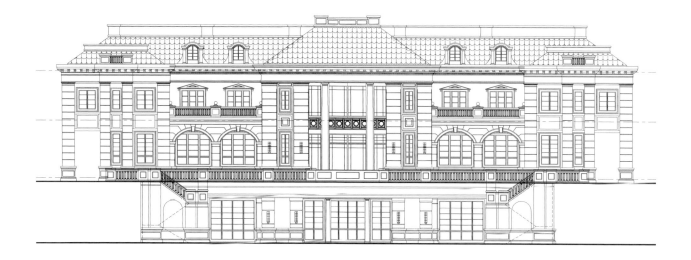

CHANGZHOU POLY PARK JIULI

常州保利公园九里

用地面积	建筑面积	设计时间	建筑类型
10.478万平方米	30.3968万平方米	2010年10月	高层住宅与别墅

本项目地处常州市钟楼区，与京杭运河老线段相望，规划分为北部高层区和南部别墅区。房型创新度领先市场水平，别墅产品演绎老城区中的法式风情，高层产品采用尊贵的Art Deco的建筑风格，让住户的尊贵感和价值感得到充分的体现。

Poly Park Jiuli, located in Zhonglou District of Changzhou City, faces to the old watercourse of Beijing-Hangzhou Grand Canal. The site is divided into north high-rise building block and south villa block. Innovation of room layout is domestically advanced. The villa block shows French-style features of old urban area and the high-rise building block demonstrates the elegant Art Deco style, providing an elegant and valuable space for dwellers.

below site plan/总平面图
right detached villa renderings/类独栋别墅效果图

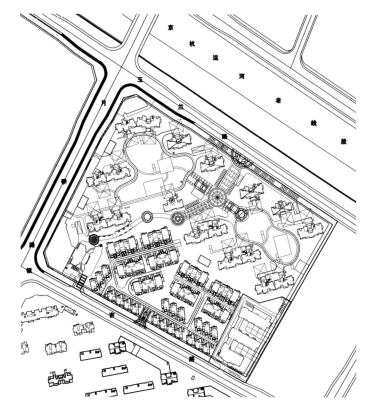

left   detached villa renderings/类独栋别墅效果图
above   detached villa elevation/类独栋别墅立面图
below   detached villa renderings/类独栋别墅效果图

left detached villa renderings/类独栋别墅效果图
below detached villa room floor plan/类独栋别墅房型平面图

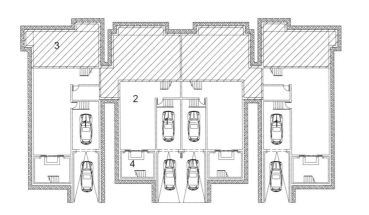
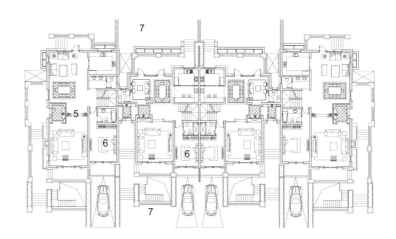
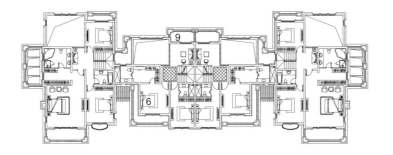
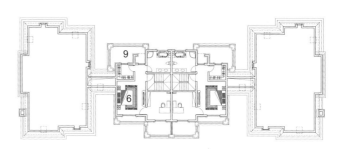

1. garage /车库
2. storage room/储物间
3. light well/采光井
4. sunk lourt/下沉庭院
5. living room/客厅
6. bedroom/卧室
7. garden/庭院
8. study/书房
9. terrace/露台

KUNSHAN 21 NEW CITY

昆山21新城

left site plan/总平面图
right club detail photo/会所细部实景照

用地面积	建筑面积	设计时间	建筑类型
约56.4万平方米	约165.7万平方米	2009年至今	商办+高层住宅+别墅

本项目基地东至洞庭湖路，南至景王路，西临吴淞江路，北至白士浦河。建设用地面积约56.4万平方米，规划总建筑面积约165.7万平方米。周边公交路线与轨道交通线较多，公共交通良好。规划采用先进的混合社区模式，通过产品混合和人群混合，综合开发用地，将居住、商业、酒吧娱乐、办公、酒店等业态整合为一个整体。通过"次街"的设置，将小区道路与居民生活进行融合，小区内人车分流，小区外人车共生，增加居民交流以及活动的空间。同时，通过高层区与低层区的合理布置，使高层享受到最大化的景观资源，同时也保证了低层区的私密性。

Kunshan 21 New City is located on a site spreading out to Dongtinghu Road at east, Jingwang Road at south, Wusongjiang Road at west and Baishipu River at north. The project covers a land area of about 564,000m² and a total planning building area of about 165,700,000m², and enjoys a convenient public traffic system consisting of many bus lines and rail lines. The planning adopts advanced multifunctional community mode and realizes a general real estate development through mixture of diversified building types and dweller classes, so as to integrate various functions, such as residence, commerce, bar, entertainment, office and hotel, etc. "Secondary street" is designed to combine the community paths into dwellers' daily life, separate pedestrain and vehicle flows in the community and realize coexistance between pedestrian and vehicle systems outside the community, so as to enlarge the spaces for dwellers' communication and activites. Meanwhile, high-rise building block and low-rise building block are reasonably arranged to enable high-rise buildings to enjoy maximum landscape resources and to protect the prviacy of low-rise building.

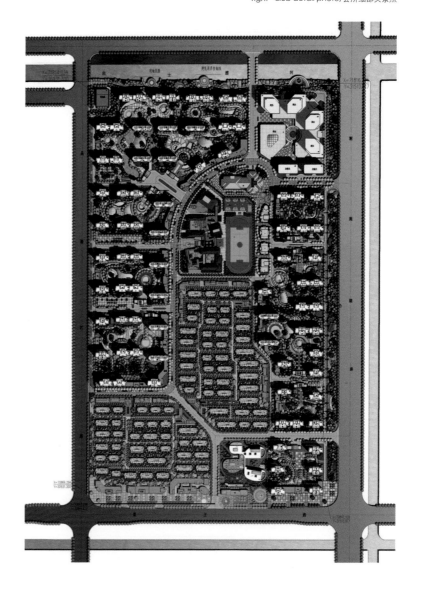

left club photos/会所实景照
right section diagram and high-rise buildings photos/节点图解和高层实景照

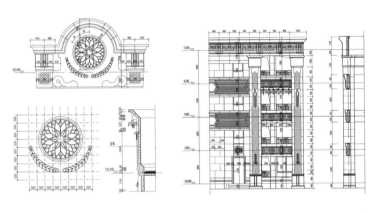

below villa exterior view/别墅外景图

below villa night view/别墅夜景图

DALIAN XINGHAO STAR RESIDENCE

大连星浩星光域

用地面积	建筑面积	设计时间	建筑类型
89,900平方米	294,069平方米	2011年1月	高端高层住宅

本案地处大连市中心城区东部，未来大连的"维多利亚港"——东港。项目为该地区仅有的18块住宅用地中的3块，是极其稀缺的海岸高端居所。规划创造性地采用"九宫布局"，将四个城市街角放开，打造绿色组团入口花园，以减少视线遮挡和噪声干扰。落车亭、风雨长廊丰富了生活空间的层次。120米超高层公寓呈南低北高对称布置，尽显庄重礼仪气质。建筑立面简洁大方，全干挂幕墙体系，融合经过提炼的细部符号，玻璃与石材幕墙虚实对比，烘托出楼盘高贵优雅的整体气质。

Dalian Xinghao Star Residence project is located at East Harbor, which will become "Victoria Harbor" of Dalian in the future, at east urban area of Dalian City. The project contains three of the 18 plots for residential purpose in this area and is a rare coastal high-end residence. The planning creatively adopts "Nine-Grid Layout" to open the four urban corners and to create a green block entrance garden, which could obviously reduce visual barricade and noise disturbance. The introduction of stop shelter and open corridor enriches the living spaces. 120m super-high apartment is designed symmetrically with the north part higher than the south part, demonstrating a solemn and elegant image. Simple but generous facades are made of dry-hung curtain wall system, in which refined detail symbols are integrated. Glass and stone curtain wall form an intensive constrast between virtuality and reality, embodying the general elegant and solemn image of the project.

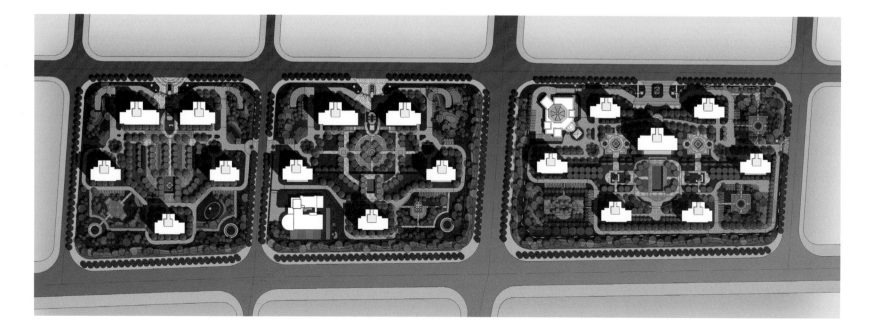

left site plan/总平面图
right exterior view/外景图

left exterior view/外景效果图
below exterior photo/外部实景照片

above exterior photo/外部实景照
left detail photo/建筑细部实景照
right interior photo/建筑内部实景照

DALIAN HONGXING SEA WORLD VIEW PLOT 14 17

大连红星海世界观14、17号地块

用地面积	建筑面积	设计时间	建筑类型
14#：28 671平方米	14#：15 297平方米	2011年8月至今	别墅及高层住宅
17#：98 491平方米	17#：111 923平方米		

　　700万平方米原生态山林环簇，4.9千米海岸线绵延，一线山海风光兼得——这就是红星海世界观。14#、17#地块作为大盘别墅产品的收官之作，设计定位为浪漫休闲，慵懒度假的托斯卡纳风格滨海高端别墅住区。规划采取自由生长的叶脉状布局形式，顺应地形精心安排道路、布置产品，通过错位、高差变化等设计手段，户户可以赏山观海。弱连接脱开相邻单元，营造出独立式建筑的居住特征。建筑造型采用纯正的意大利托斯卡纳风格，省略繁复的雕琢和装饰，用简单、圆润的线条，浑圆的修边营造出返璞归真的感觉。建筑细节精雕细琢，手工打造的天然石材外立面，高低错落、色彩不均的筒瓦屋顶，爬满藤蔓的壁炉，壁灯点亮的庭院，花样铁艺，汩汩的喷泉……掩映在碧天蓝海之下，营造出充满历史感和欧式风情的居住氛围，极大地扩展了现代居住体验，提升了别墅自身的价值。

Hongxing Sea World View covers 7,000,000m² of original ecological and mountainous environment and is hugged by 4.9km coastal line, enjoying beautiful mountainous and waterside landscape. Plots 14 and 17 as the final segment of large-scale villa project are designed as high-end holiday villa zone of Tuscany style with romantic features. The planning adopts a layout of freely growing leaf vein and paths are designed carefully according to terrain condition. Mountainous landscape and waterscape are introduced into each room through stagger layout and different heights of buildings. Adjacent units are relatively independent but also connected at some positions to provide a unique dwelling experience in independent residence. Buildings adopt pure Italian Tuscany style to omit complicated decoration and finishing and to create an original and natural appearance through simple and smooth lines and edges. Architectural details, such as facades made of natural stone, pantile roof with different heights and colors, fireplace covered by green vine, courtyard lighted by wall lamp, iron pattern and fountain reflected under azure sky and on blue sea are designed precisely to create a residential space full of historical and European features, enhancing the experiences of modern residence and improving the value of villa building.

left sketch bird's eye view/手绘鸟瞰图
right renderings/效果图

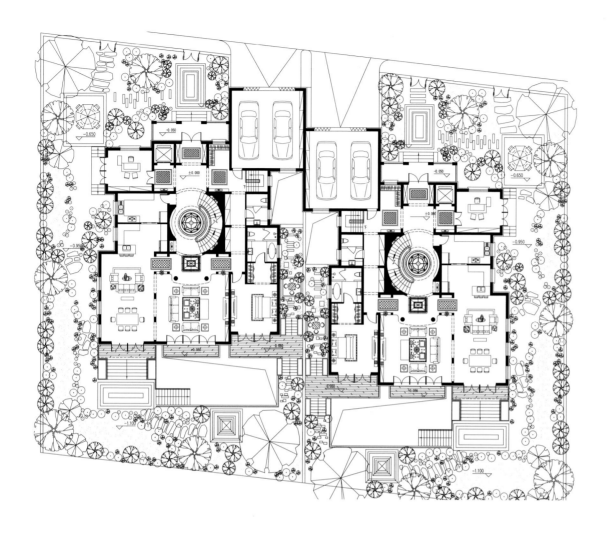

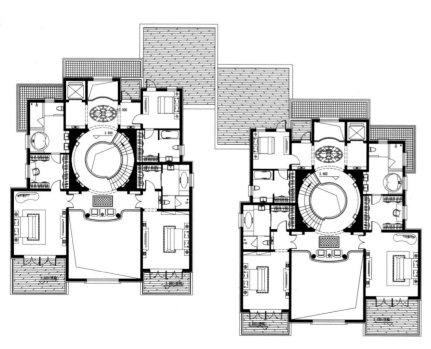

top left semi-detached first floor plan/双拼一层平面图
bottom left semi-detached second floor plan/双拼二层平面图
right semi-detached renderings/双拼效果图

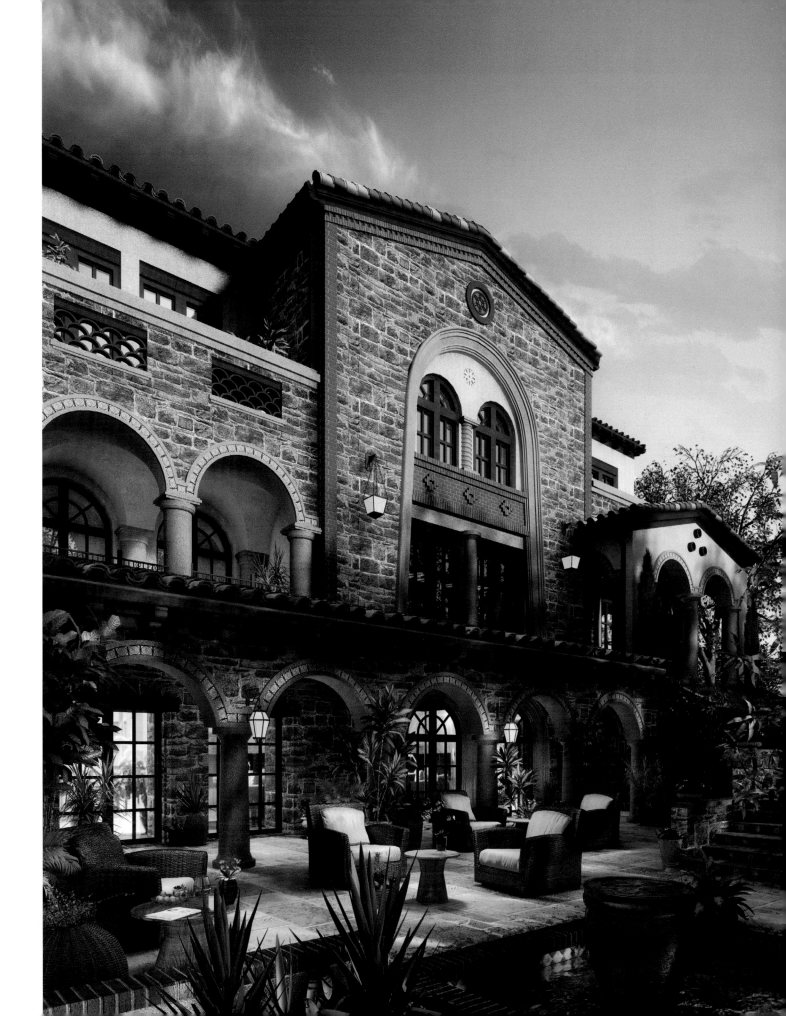

DALIAN HAICHANG AUTO CITY RESIDENTIAL COMMUNITY

below site plan/总平面图
right flavor commercial street/风情商业街

大连海昌汽车城居住区项目

用地面积	建筑面积	设计时间	建筑类型
275 890.9平方米	306 404.8平方米	2011年5月至今	住宅

本项目在大连市保税区二十里堡这一方热土上，以莎士比亚故乡斯特拉福德镇为摹本，在红砖墙、矮烟囱、老虎窗等鲜明古镇符号的映衬下，打造了一个俊朗、幽静、休闲的英伦风情小镇。

"一条景观轴线，两片英伦特色商业街"整合了整个基地，并将周边的高尔夫别墅区、汽车城风情商业广场、浓郁的山脉景观融入本小镇之中，打造了原汁原味的英伦旅游度假小镇。

漫步在古朴的青石板商业小街上，漫步在葱翠精致的别墅庭院前，仿佛回到了16世纪的纯朴风景与别致造型，郁郁葱葱的草坪和花木映衬着色彩鲜艳的红墙、白墙、黑瓦，显得优雅、庄重。幽深碧绿的藤蔓正努力向两旁的红砖墙上攀爬，抬头看到那座露台，仿佛朱丽叶正在那里喃喃叙述自己对罗密欧的绵绵情意，一段传颂千古的爱恋悠然展开。

Dalian Haichang Auto City Residential community is located on Ershilipu Street in the Bonded Area of Dalian and is a simulation of Stratford town where is the hometown of Shakespeare. Eye-catching symbols of ancient town, such as red brick wall, short chimney and dormer, successfully create a beautiful, peaceful and cozy British-style town.

"One landscape axis and two British-style commercial streets" are designed on the project site to integrate surrounding golf villa area, auto city flavor commercial plaza and beautiful mountainous landscape into this small town, creating an original British-style residential town for tourism and holiday.

Walking on the commercial streets paved by ancient blue slates and on the paths in front of villa courtyard, people will feel standing among pure natural landscape and unique buildings in the 16th century. Green lawn and trees and colorful flowers make dialogue with red and white walls and black tiles, producing a kind of elegant and solemn atmosphere. Verdurous vines are striving to grow on the red brick walls on both sides. The overhead terrace gives people a feeling just like Juliet is opening her heart to Romeo and a romantic story is happening.

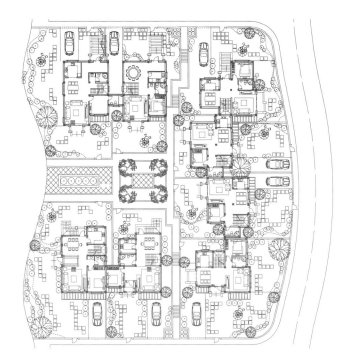

below   exterior view of villas/别墅外景图
above   courtyard villa 1st floor plan/合院别墅一层平面图

SHENYANG GREENLAND STATE GUEST HOUSE

沈阳绿地国宾府

用地面积	建筑面积	设计时间	建筑类型
26.7万平方米	22.5万平方米	2011年至今	低层住宅

本项目目作为辽宁省国际会议中心建设项目的配套居住部分，拥有沈阳最优质的自然资源和人文资源，坐拥山景、水景及高尔夫球场等多种自然景观。设计利用优越的区域环境，将山、水、高尔夫球场及简约的法式风情融为一体，营造休闲而养生的生活空间，打造"自然的、生态的、健康的、休闲的"上善居品。建筑群体强调与自然地形的结合，充分利用原始地形的自然形态，灵活布置建筑群体，使其错落有致、空间序列变化丰富。同时总体布局深入关注与大地轮廓线的共生，并自然生发出丰富的内部公共空间，使建筑与自然环境有效地融合在一起。

Shenyang Qipanshan Residence, as the supporting residential area of Liaoning International Conference Center, enjoys the best natural resources and humanistic culture of Shenyang, as well as diversified natural landscapes, such as mountainous landscape, waterscape and golf course, etc. The design utilizes advantageous regional conditions to integrate mountaineous landscape, waterscape and golf course into the simple French-style residence, successfully creating a cozy and healthy living space with "natural, ecological, healthy and leisure" features. With consideration to natural terrain conditions, buildings are flexibly staggered with each other to realize diversified spatial layout. Meanwhile, coexistence between the general layout and ground contour line is emphasized carefully and abundant internal public spaces are formed consequently to effectively integrate buildings with natural environment.

above site plan/总平面图
right exterior photo/建筑外部实景照

left and right exterior photos/建筑外部实景照

above   club interior photo/会所室内实景照
right   villa interior photos/别墅室内实景照

CHANGCHUN XINLI CENTRAL KUNGKUAN

长春新里中央公馆

用地面积	建筑面积	设计时间	建筑类型
228 658平方米	725 797.61平方米	2007年至2010年12月	商业+办公+住宅

长春中央公馆位于吉林省长春市南关区，基地北侧为城市环形干道南环路，东南紧邻幸福街，西侧为南湖中街，基地为三角形，内部另有两条城市支路通过，将基地划分为A、B、C三个地块。本项目设计中首次引入创新的"三级物管"模式，引入"公馆"概念。在小区的入口处，设置入口大堂，增强小区的私密性。大堂之后即为内院，内院连接至各单元入户大堂，更增加了小区居民的归属感和安全感。

Changchun Central Kungkuan, located in Nanguan District of Changchun City, Jilin Province, is adjacent to Nanhuan Road which is an urban ring trunk road at north, Xingfu Street at southeast and Nanhu Middle Street at west. The project is built on a triangular site, where two pieces of urban branch roads pass through, dividing the site into blocks A, B and C. Planning design of this project adopts innovative "three-grade real estate management" mode and "Kungkuan" (resident of a rich or an important person) concept for the first time. A lobby is designed at the community entrance to protect community privacy. A court is arranged behind the lobby and is connected to the entrance hall of each unit, further enhancing residents' sense of belonging and security.

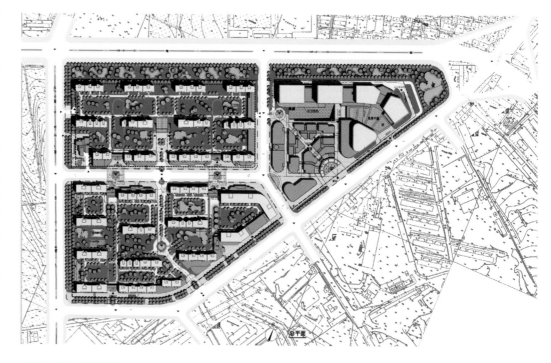

above site plan/总平面图
right group inner courtyard photo/组团内院实景照

left residential photo/住宅实景照
top right residence hall interior view/入户大堂内景图
bottom right residence hall exterion view/入户大堂外景图

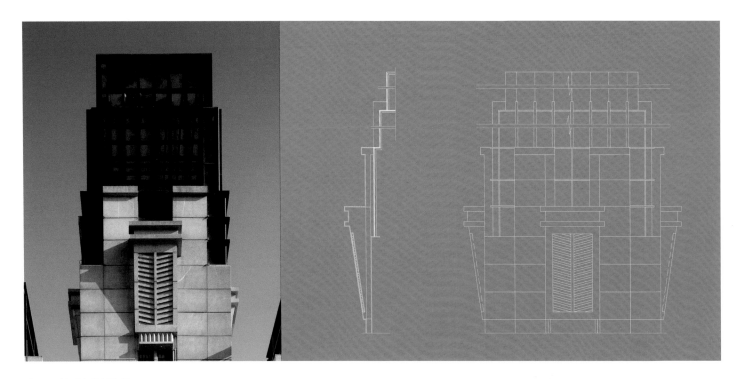

above　club detail/会所细部
below　club facade/会所立面
right　club photo/会所实景照

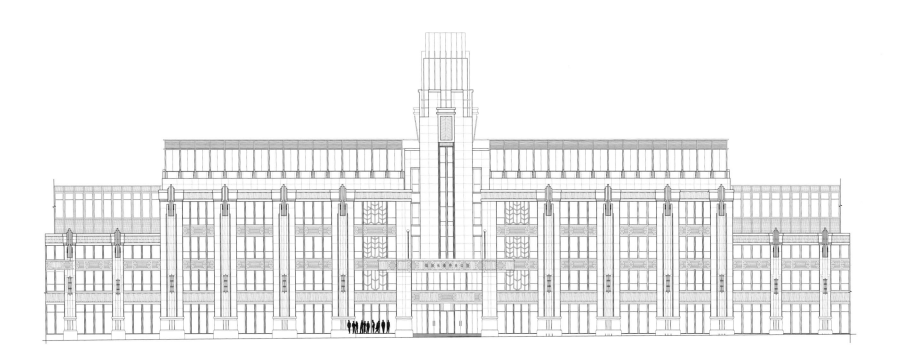

GREENLANDNATHAN MANOR

绿地内森庄园

用地面积	建筑面积	设计时间	建筑类型
141 149平方米	271 560平方米	2009年至2010年7月	豪宅

作为绿地集团在合肥市唯一的庄园级府邸，内森庄园拥有城市中不可复制的景观资源—天鹅湖、中央河、防护林景观带。除拥有上佳景观资源外，内森庄园还位于合肥市政务新区黄金地段，设计师希望将之打造为"健康、舒适、环保的一体化产品。"其中包括实体产品、物业服务、高端配套等。

整体规划采用高层区和低层区的明确分区，互不干扰，有利于物业的分开管理。高层区位于总地块的北面，设置了独立的出入口和各级物业管理，同时创新采用了"底层交通系统"，利用架空层创造出宜人的回家路径和泛会所，底层系统及泛会所采用室外设计室内化的装饰原则，给人以富丽堂皇的感受。相对于高层区的统一物业管理，别墅区的规划显得更私密、更自由，重点营造庭院深深、世外桃源般的景观体验，体现低调、内敛的风格特征；高层区布局，重点控制轴线的设置，建筑与景观序列的组合关系与尺度配合，全面彰显空间的庄重与尊贵。

Nathan Manor, as the unique manor-level mansion developed by Greenland Group in Hefei City, enjoys the best urban landscape resources, such as Swan Lake, Central River and protection forest landscape belt, etc. Besides excellent landscape resources, this project is located at a golden area in New Municipal and Cultural District of Hefei City. The designers hope to create Nathan Manor into a "healthy, comfortable and environmentally-friendly" integrated project, including real estate products, real estate management service and high-end supporting facilities, etc.

General planning of this project clearly sets high-rise buildings and low-rise buildings in two independent blocks without mutual disturbance, so as to be convenient for separate real estate management. High-rise building block is located at the north part of the site and is designed with independent entrance and all levels of real estate management. Meanwhile, innovative "ground floor traffic system" is adopted to design a convenient path guiding home and outdoor chamber on the overhead layer. Ground floor and outdoor chamber adopt interior-type outdoor decoration, realizing a palatial image. Compared with the unified real estate management of the high-rise building block, the planning of villa block becomes more private and free, focusing on creating deep court and beautiful landscape and demonstrating modest and introversive characteristics; planning layout of high-rise building block emphasizes on the design of axial line, relationship between building and landscape and scale proportion, so as to show solemnity and elegance of the space.

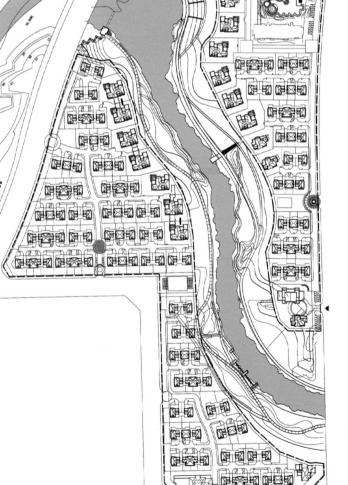

above bird's eye view/鸟瞰实景图
below bird's eye view/鸟瞰图

left   villa photos/别墅实景照
above   club perspective/会所透视图
below   elevation of club/会所立面

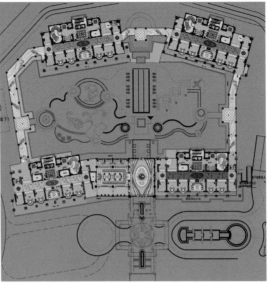

left and right club ground floor system and photos/
泛会所底层系统和实景照

HEFEI BINHU TIME INTERNATIONAL PLAZA

合肥滨湖时代国际广场

用地面积	建筑面积	设计时间	建筑类型
11.37万平方米	46.76万平方米	2010年9月至今	高层住宅

项目位于合肥市滨湖新区，东临徽州大道（南北向主干道），南临杭州路，西临西藏路，北临烟墩路。

整个地块为一较规则长方形用地，地块东区为办公及商业，西区为住宅，两个区域相互独立，又融为一体。住宅区域分为南北两个组团，分别在南北城市道路上设置组团出入口，同时共享两个组团间的大型集中绿化。整体立面造型引入新古典的比例控制思想，立面以"三段式"构成为要素，在构图和均衡方面均细加推敲以至完美，注重建筑顶部的灯光设计，使得每一个建筑显得端庄、沉稳，营造出优雅统一又不失温馨浪漫的新古典风情社区氛围。

Binhu Time International Plaza, located in the Binhu New District of Hefei City, is adjacent to Huizhou Highway (a south-north trunk road) at east, Hangzhou Road at south, Xizang Road at west and Yandun Road at north.

The site is regularly rectangular. Office and commercial buildings are designed at the east part and residential buildings at the west part. These two parts are relatively independent but also mutually connected. The west part is divided into south and north blocks, whose entrances are set on the south and north urban roads respectively. These two blocks share a large central green space bewteen them. Neoclassical proportional control concept is introduced into facade design and "three-segment" structural elements are used to carefully design structural pattern and equilibrium, so as to realize perfect effect. Light design on building roof is emphasized to produce elegant and solemn building appearance and create a generous, uniform, cozy and romantic neoclassical community atmosphere.

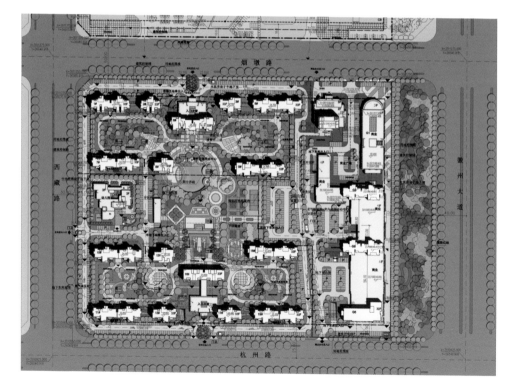

above site plan/总平面图
right exterior view/外景图

left and right   architectural exterior view/建筑外景图

HEFEI INTERNATIONAL FLOWER CITY DAISY PARK

合肥国际花都·蓝蝶苑

below site plan/总平面图
right exterior view/外景图

用地面积	建筑面积	设计时间	建筑类型
32 402平方米	22 013平方米	2009年	教育配套

合肥国际花都·蓝蝶苑位于合肥市政务新区。以高层产品为主。住宅组团尺度宜人，内部空间疏朗有序。建筑造型采用Art Deco风格，运用简洁流畅的线条，强化住宅挺拔向上的形态气质。

Hefei International Flower City Daisy Park, located in New Municipal and Cultural District of Hefei City, is a high-rise building project. Residential building block has an appropriate scale and regular internal spaces with proper density. Building shape adopts Art Deco style, simple and smooth lines are used to enhance the ascending image of these residential buildings.

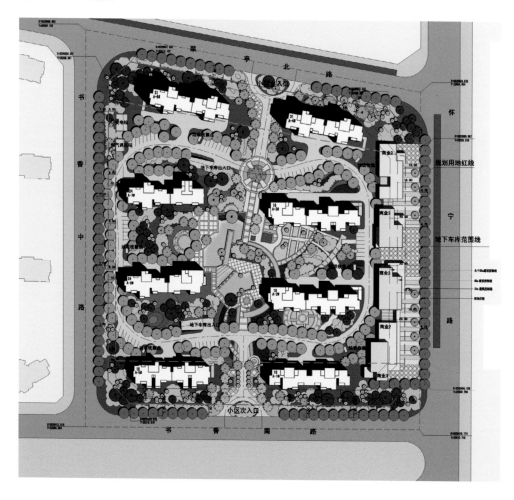

left and right high-rise uildings exterior view/高层外景图

XI'AN HAIYU NOTINGHILL VILLA

西安海域诺丁山

用地面积	建筑面积	设计时间	建筑类型
54 722.81平方米	82 234.2平方米	2008年	住宅

项目基地位于西安市高新区CBD北侧，南临10万平方米城市公园。规划以全景别墅理念为出发点，整体大地库设计使每套别墅均拥有直通地下层的专用车库；同时使地面实现全铺装景观设计，营造出安全宜人的大花园式高档别墅区。建筑采用极具风情的英式风格，红色系的仿古砖与周边绿化及城市公园交相辉映，大大舒缓了CBD核心区密集的现代化高层和超高层建筑形成的压抑感，创造了一处难得的高品质都市公园别墅区。

Haiyu Notinghill Villa is located at north of CBD of Xi'an High-Tech Zone and is adjacent to a urban park covering an area of 100,000 m^2 at south. The planning is based on the concept of panoramic villa and the large basement design provides underground private parking lot for each villa building; meanwhile, the ground is paved and designed with landscape, creating a safe and cozy large garden-type high-end villa community. Buildings are designed with British-style features. Red antique brick is harmonious with surrounding vegetation and urban park landscape, alleviating the pressure caused by dense modern high-rise buildings and super-high buildings in CBD core area and successfully creating a rare high-quality urban park villa zone.

left site plan/总平面图
right exterior view/外景图

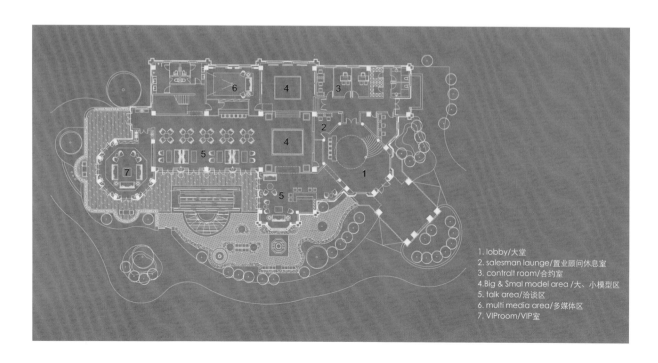

1. lobby/大堂
2. salesman launge/置业顾问休息室
3. contralt room/合约室
4. Big & Smal model area /大、小模型区
5. talk area/洽谈区
6. multi media area/多媒体区
7. VIProom/VIP室

bottom left   club plan/会所平面图
top left and right   exterior view/外景图

left   exterior detail/外景细部
right   exterior view/外景图

GREENLAND SAINT LOUIS PALACE

绿地·圣路易名邸

用地面积	建筑面积	设计时间	建筑类型
80 027.71平方米	318 244.9平方米	2010年至2011年1月	住宅

绿地·圣路易名邸位于成都市城西板块，以"成都住区样板区"著称。距离成都三环路仅8分钟，距离成都市中心约为9千米，交通便捷。

地块规划总净用地面积80027.71平方米，总建筑面积318244.9平方米.其中地上建筑面积222389.74平方米，容积率2.80，建筑密度0.24。

规划上采用大围合&大花园、立体人车分行、三级物业管理模式。整体风格体现法国古典文化，旨在创造高档居住社区，为城市添彩，为业主创造高雅浪漫的生活环境，以此来引导一种高尚的生活方式，并将这一理念贯穿于设计始终。

Saint Louis Palace is located in the west part of Chengdu city, familiar as "the sample residence in Chengdu city". 8 minutes' drive from the third Urban Circle Arterial Street in Chengdu city, 9 kilometers' distance from the center of the Chengdu city, it is convenient in traffic.

The project covers a total net land area of 80027.71㎡, with a total built-up area of 318244.9㎡ and a total built-up area of 222389.74㎡ above ground, gaining a plot ratio of for 2.80 and a construction density of 0.24.

The plan adopts the manage ment mode of constructing it with both big combinations and great gardens, dividing the vehicle flows in three-dimensional space and managing the properties at third-level. The whole style of the project is sunken in a taste of France classical culture, aiming at establishing a noble community, blowing a new air in city, creating a grace and romantic environment, which will lead an elegant living style advocated throughout the designing.

below bird's eye view/鸟瞰效果图
right residential photo/住宅实景照

left and right residential photo/住宅实景照

5 城市规划与城市设计
City Planning and Urban Design

随着中国城市化的快速推进，不断扩张的城市亟须规划整合和城市设计的控制。在理性梳理城市资源的基础上，优化城市运行的几大系统，并在城市成长过程中提升土地价值和城市实力，创造理想的空间舞台始终是UA规划的目标。依据理性的空间句法，从城市设计维度着力于城市公共空间系统和配套设施系统的设计，必将创造出具有场所感和归属感的个性鲜明的现代宜居城市。同时，生态、低碳等符合国际潮流的先进理念是UA城市规划与城市设计的永恒维度。

With acceleration of urbanization in China, ever-expanding urban areas urgently require control on integration planning and urban design. Based on rational analysis of urban resources, UA takes its constant planning objectives to optimize the large systems for urban operation, improve land value and urban competitiveness during urban development and create ideal spaces. According to rational spatial design principle, the urban design focusing on realizing urban public space system and supporting facilities will definitely create a modern and habitable city providing strong sense of space and belonging and demonstrating outstanding features. Meanwhile, advanced design concepts, such as ecological protection and low carbon discharge, which meet international development requirements, are UA's constant aim in urban planning and design.

TAI'AN TAISHAN MOUNTAIN WEST PIEDMONT

泰安泰山西麓项目

用地面积	建筑面积	设计时间	建筑类型
355.78万平方米	108.32万平方米	2011年10月	商业、商务会展、旅游文化、住宅

泰山，中华文明的重要发祥地之一，五岳之首，华夏文化的缩影。本案规划总用地约5 000亩，位于泰山西麓，泰安中心城区西北面。通过对区域开发模式、产品架构、发展途径、空间布局等方面的综合审视，提出了以打造绿色"森林家（Trees Home）"为核心的规划理念。项目以泰安市城市总体规划为指引，深入挖掘和大力弘扬泰山文化，积极推进旅游和文化产业融合，着力打造国内顶级、国际一流、集"旅游文化、商务会展、特色商业、生态宜居"产业为一体的绿色城邦。

Taishan Mountain is one of the important cradles of Chinese civilization and is ranked first among five famous mountains in China, showing a panorama of Chinese culture. This project, covering a total planning area of about 5000MU (MU is a unit of area and 1MU is about 667m²), is located at west foot of Taishan Mountain and northwest of urban center of Taian City. After general consideration of regional development mode, product planning, development approach and spatial layout, the core planning concept is formed to create green "Trees Home". The project is based on the general urban planning of Taian to explore and propagandize the culture of Taishan Mountain in depth, positively promote integration of tourism and cultural industry and make great efforts to create a domestically supreme and internally advanced green city integrating tourism culture, business exhibition, featured commerce and ecological residence.

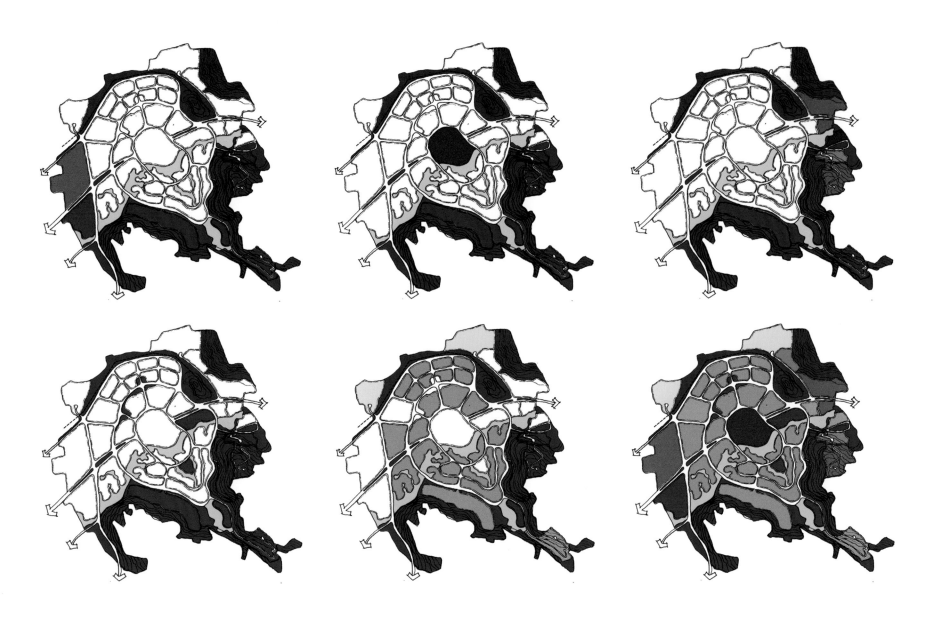

left program distribution diagram/方案功能图解
right site plan/总平面图

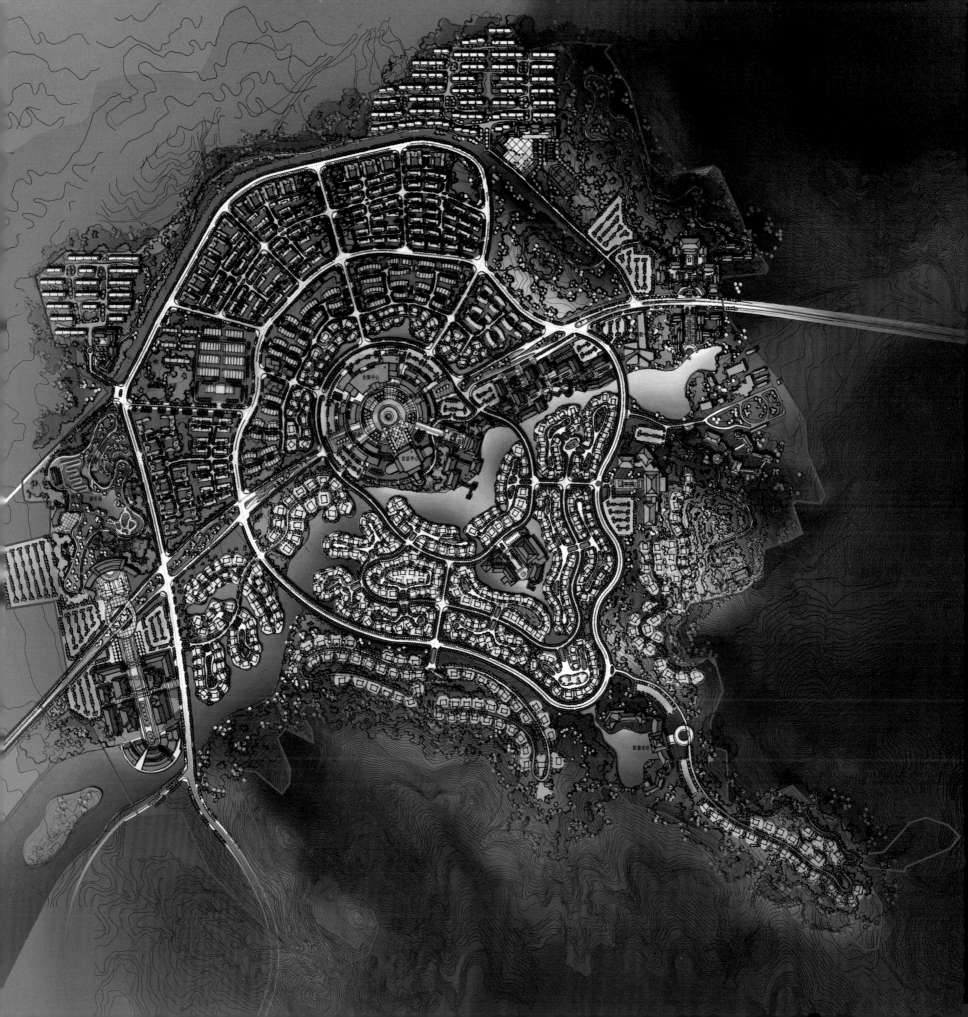

KAIFENG BIANXI NEW DISTRICT CONCEPT PLANNING

开封汴西新区概念规划

用地面积	建筑面积	设计时间	建筑类型
778 303平方米	2470 935平方米	2011年11月至2012年1月	商务办公及住宅

　　开封汴西新城项目地处开封市汴西新区，是开封郑汴一体化发展进程的重要方向。本项目是汴西新区的商务中心核，与行政中心配套构成新区的完整内容。方案设计以新区"宋式七园林"的绿地景观体系为规划基调，结合开封历史文化名城背景，将重现"清明上河图"历史胜景为规划愿景，打造适宜新区发展的商务中心、发展核心及宜居中心。

Kaifeng Bianxi New City project, located in Bianxi New District of Kaifeng City, leads an important direction for integrative development of Zhengzhou City and kaifeng City. This project is the business center of Bianxi New District and it combines with the administrative center to form a complete new district. The scheme design is based on the green landscape system of "Seven Song Dynasty Gardens" and its planning vision is to reproduce the famous historical scene described on the painting of "Riverside Fair in the Pure Brightness Festival", to create a business center, development center and habitable center in the new district.

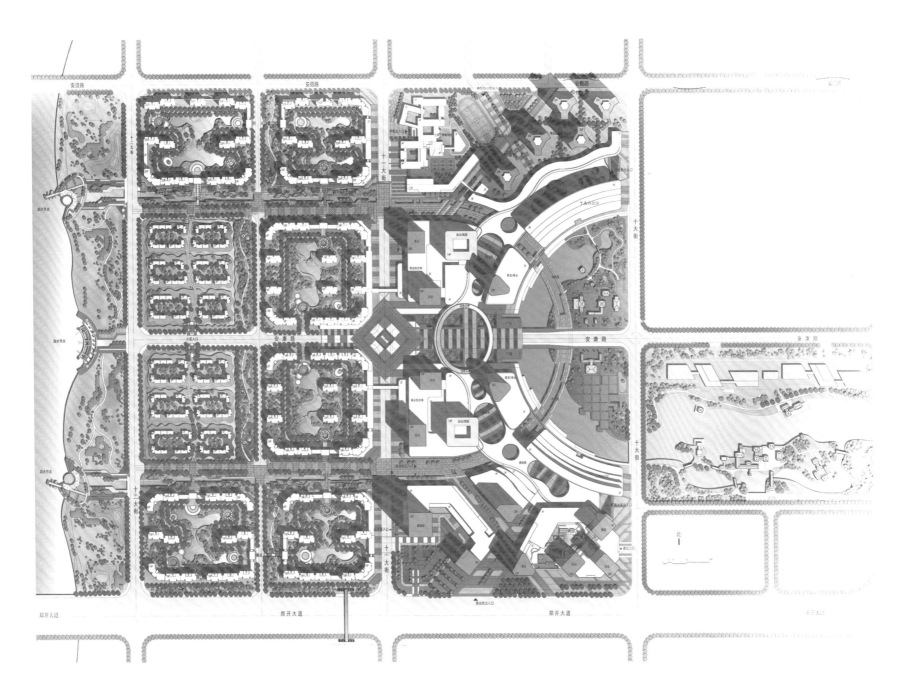

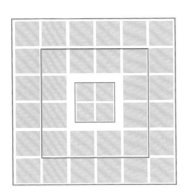
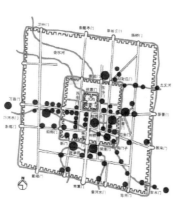
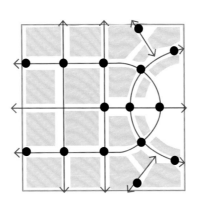

above site plan/总平面图
left bottom A floor standard planar graph / A楼楼层标准平面图
middle bottom B floor standard planar graph / B楼楼层标准平面图
right bottom modern structure planning / 现代规划结构

above bird's eye view from east sight / 东面鸟瞰图
right bird's view from west sight / 西面鸟瞰图
below concept diagram/ 概念图解

DALIAN POLY FISHERMAN'S WHARF

大连保利渔人码头项目

用地面积	建筑面积	设计时间	建筑类型
64.98万平方米	241.488万平方米	2010年6月	办公及商业综合体

大连保利渔人码头项目位于大连市中心城区内仅有的一处待开发的"钻石港湾"——东港，是大连唯一一个以"游艇码头"为主题的高品质大型综合社区，总建筑面积92万平方米。

区域中保留U形内河港池，用以停泊游艇，蔚蓝的水面映衬着白色的帆船，构成一道亮丽的水岸风景。

贯穿整个场地的中心主轴，把城市公共交通（地铁）、配套商业、游艇码头以及半公共的社区花园、私密的高档别墅区生态步行道有机联系起来，打造一条充满活力的城市生活主线。广场景观设计与周边的建筑和水域浑然一体，将人的活动从交通枢纽、商务办公自然引向水边。广场的界面、高差的处理、小品的设置使滨水空间变得丰富起来。

大连保利渔人码头成为未来滨港新城开发的范本。

Dalian Poly Fisherman's Wharf project, located at East Harbor which is an undeveloped "diamond harbor" at the central area of Dalian City, is the unique high-quality large complex community with the theme of "yacht wharf" in Dalian, covering a total building area of 920,000m².

The U-type inner harbor is reserved for yacht docking. White sailing boats are reflected on the clear cerulean water surface, creating beautiful waterside landscape.

The main central axis passing through the whole site organically connects urban public traffic (subway), supporting commerce, yacht wharf, semi-public community garden and ecological pedestrain path at private high-end villa area together, forming a main vigorous route for urban life. Square landscape design is integrated with surrounding buildings and waterscape, naturally turning people's activites from traffic hub and busness office to the waterside. The waterside space is enriched by boundary treatment and different heights of the square and configuration of small landscape elements.

This project has become a demonstration project for development of Bingang New City in the future.

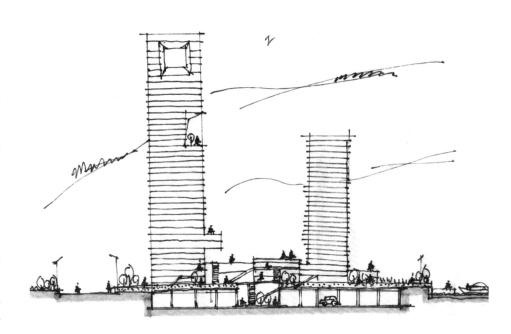

above   sketch of vertical design/竖向分析手绘图
right   bird's eye view/鸟瞰图

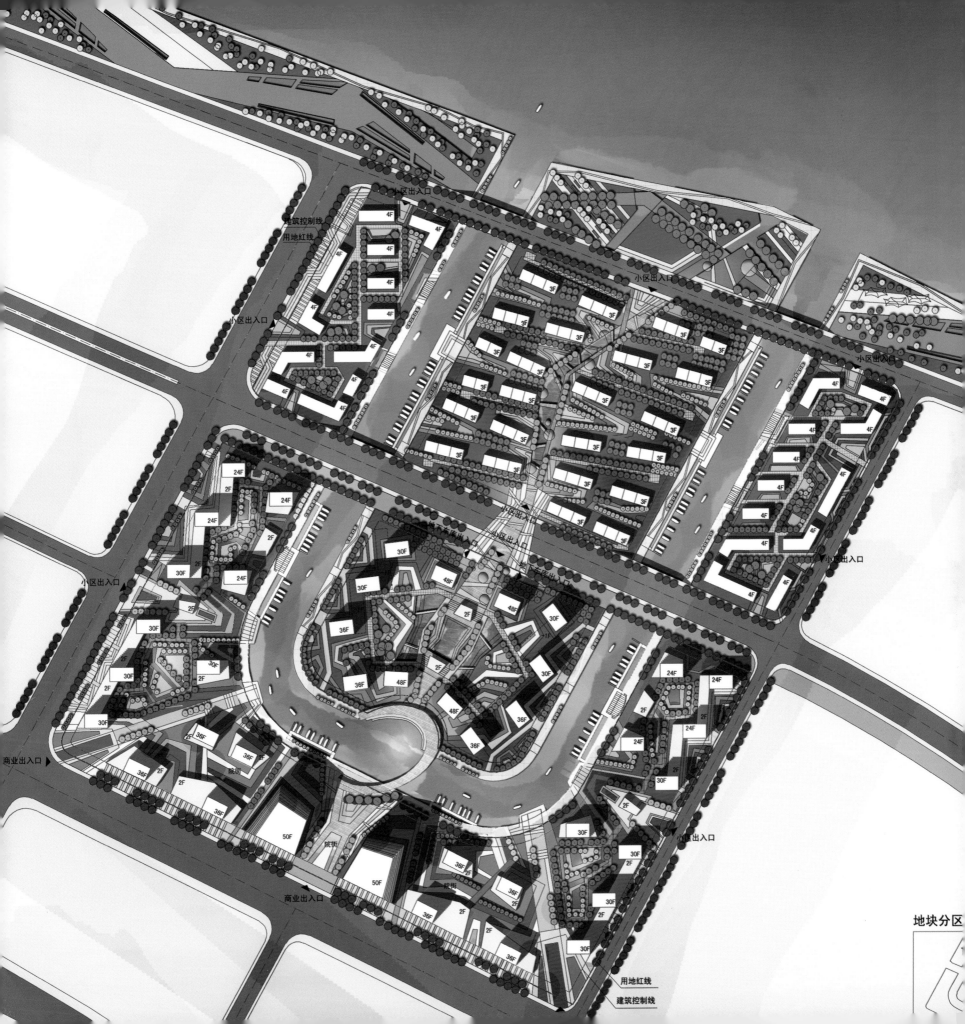

below design diagram/设计图解
right bird's eye view/鸟瞰图

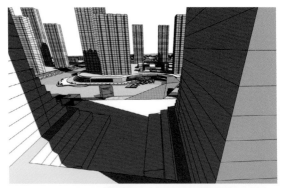
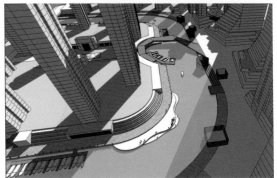
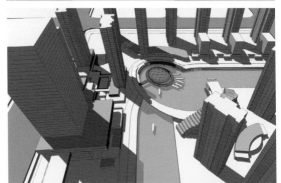
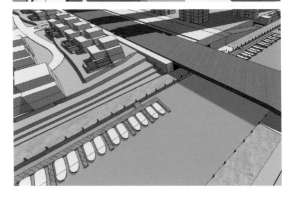

XI'AN INTERNATIONAL ECOLOGICAL CITY

西安国际生态城

用地面积	建筑面积	设计时间	建筑类型
2 873 000平方米	1 980 000平方米	2011年1月至今	混合开发

中国的根在西安,西安的肺在浐灞。浐灞生态区的整体规划强调生态网络的有机生长,注重规划景观的序列格局,是第三代新城的代表。本项目作为浐灞生态区的南翼收官之作,以"双轴双核两圈层"为主体结构,着力打造"ITEC"——国际化游憩生态新城。公建区主要由垂直花园都市、风情滨水商业街、庭园低碳办公、国际教育园区组成;居住区方面,以五大策略体现生态和活力的宗旨,用真正可持续的节能技术措施打造国际生态新城,构建西安"绿"生活。

Xi'an enjoys profound historical culture and Chanba Ecological District is an key node for urban development of Xi'an. The overall planning of Chanba Ecological District focuses on organic growth of ecological net and regular landscape layout. it is a representative district for the third generation of new city. This project, as the final segment of south wing of Chanba Ecological District, adopts a overall layout of "double axes, double cores and dual rings" and aims to create an international tourism ecological city (ITEC). The public block consists of vertical urban garden, flavor riverside commercial street, low-carbon courtyard business and international education park; the residential block expresses ecological and vigorous characteristics through five strategies and uses truely sustainable energy-saving technical measures to create an international ecological new city and realize "green" life in Xi'an.

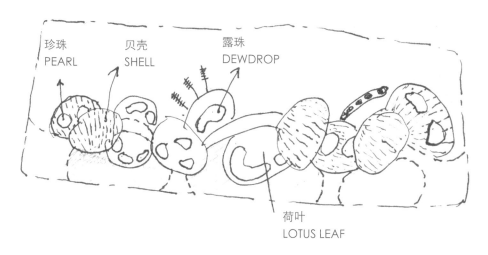

left   concept analysis/概念分析
right   site plan/总平面图

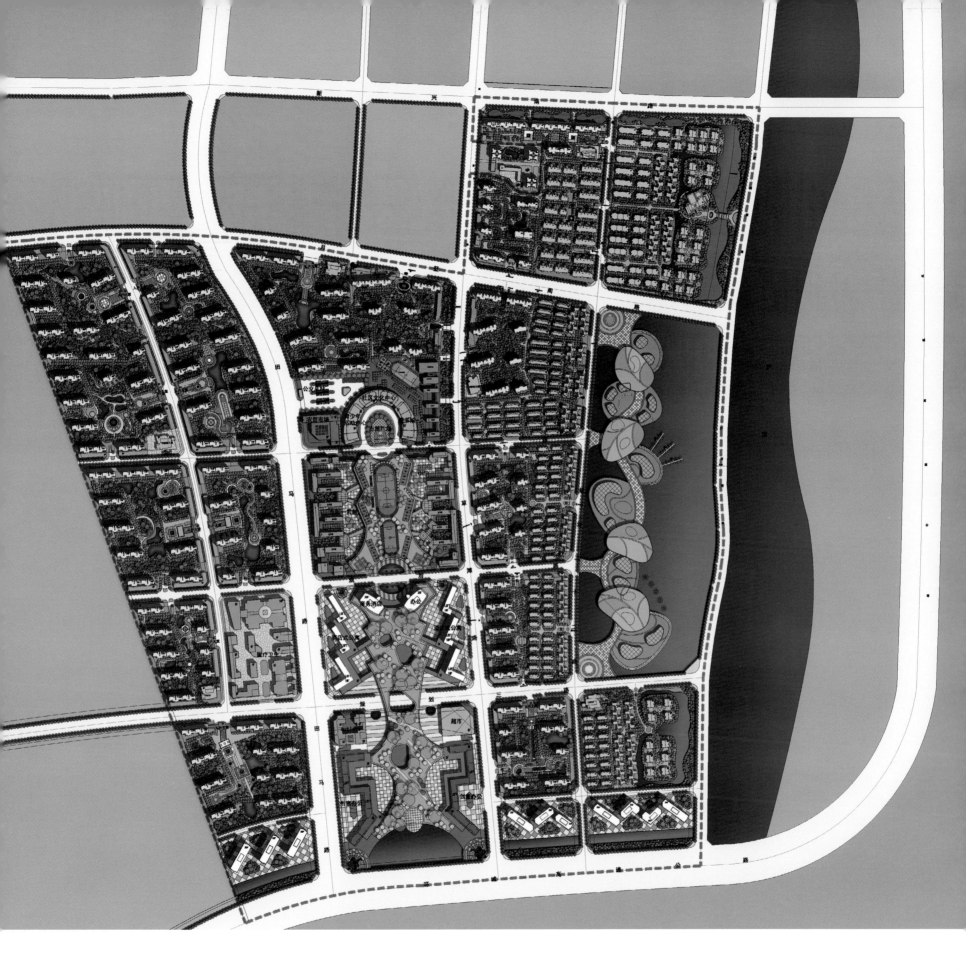

left concept analysis/概念分析
right ecological axis bird's eye view/生态绿轴鸟瞰图

VRVMCHI MOVNTAIN ZONE CONCEPTUAL PLANNING

乌鲁木齐红光山片区概念规划

用地面积	建筑面积	设计时间	建筑类型
73.16万平方米	201.13万平方米	2011年6月	商办酒店居住

新疆乌鲁木齐国际会展中心片区是未来欧亚经济贸易交流的核心地区。项目整体规划依托区域特征，力求打造国际生态商务港。倡导以大型公共设施、生态景观环境、公共交通出行为引导的发展模式。提出全新的"Best City"理念，打造完美都市。规划灵感来源于"金蚕吐丝、化茧成蝶"的概念，将会展中心喻为新丝绸之源，汇集资源，辐射周边地块，在项目地块内进行孵化，孕育新思维。规划布局以"蚕""丝""茧""蝶"作为功能区块划分及建筑形态意向的依据，功能与形态完美融合，打造集居住、办公、商务、出行、文娱、社交、休憩多功能的"Hopsca"。

Urumchi International Exhibition Center Zone is a core area for Europe-Asia economic trade and communication in the future. General planning of this project takes the regional advantages to create an international ecological business harbor. It promotes the development mode focusing on construction of large public facilities, ecological landscape environment and public traffic system. The brand-new concept of "Best City" is put forward to create a perfect city. The planning scheme is inspired from the concept of "silkworm makes silk, wraps itself in a silky cocoon and then evolves into a moth" and designs the exhibition center as a source of new silk to collect resources and connect surrounding plots, and then to hatch on this project site and generate new ideas. The planning layout takes the concept of "silkworm", "silk", "cocoon" and "moth" to divide different functional blocks and determine building shape. Function and shape are perfectly integrated together to create a multifunctional HOPSCA consisting of residence, office, business, tourism, entertainment, social communication and leisure.

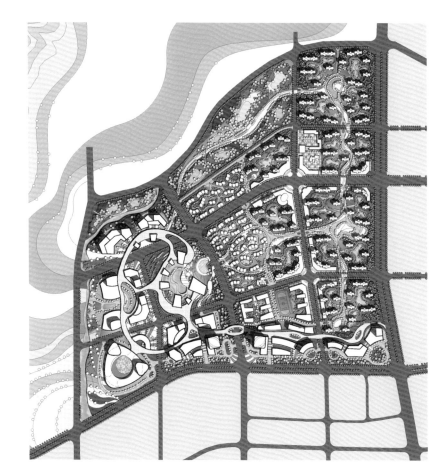

left site plan/总平面图
right bird's eye view/鸟瞰图

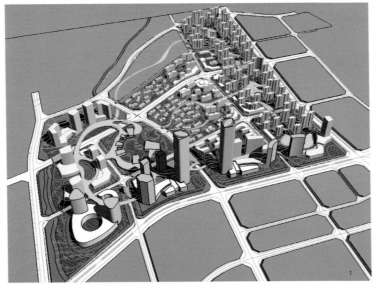

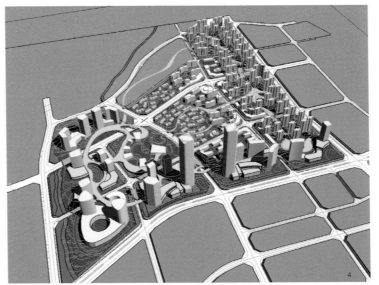

1. business/商务办公
2. green land/生态绿核
3. hotel, serviced apartment/酒店、酒店式公寓
4. commerce/商业
5. community/社区

left program distribution diagram/方案功能图解
right exterior view/外景图

GUIYANG HUAGUOYUAN

贵阳花果园

用地面积	建筑面积	设计时间	建筑类型
977 602.93平方米	7 614 500平方米	2009年12月至今	商业 办公 住宅

彭家湾·TRBD位于贵阳市彭家湾西北侧，贵黄路和川黔铁路北侧，东接松山路，西接双龙路，临近西侧的五里冲水果批发市场，南面可接花溪大道北段，花果园大街贯穿整个基地。地块规划总净用地面积977 602 .93平方米。其中地上建筑面积654 6952平方米，地下建筑面积106 7548平方米，容积率5.6。彭家湾·TRBD规划上由三个概念融合而成— TBD+CRD+CBD，意为"都市游憩商业住区"，将中央商务区、城市游憩商业区、商业住区结合起来，力图打造一个复合型的城市综合体，成为贵阳的城市新核心。

Pengjiawan TRBD is located at northwest of Pengjiawan in Guiyang City and north of Guihuang Road and Sichuan-Guizhou railway line, connected to Songshan Road at east and Shuanglong Road at west, adjacent to Wulichong fruit wholesale market at west and north segment of Huaxi highway. Huaguoyuan avenue passes through the whole site. The planning site covers a total net land area of 977,602.93㎡, among which overground building area is 6,546,952㎡; underground building area is 1,067,548㎡, realizing a plot ratio as high as 5.6. Pengjiawan TRBD planning combines three concepts, including TBD+CRD+CBD, which means that "urban recreational business residential area" integrates central business district, tourism business district and commercial and residential district together to create a composite urban complex which will become a new urban center in Guiyang.

above   bird's eye view/鸟瞰图
right   site plan/总平面图

left city space/城市空间
right riverside perspective/沿河透视图

NEW GITY RIYUE ISLAND CENTRAL CULTURAL AREA URBAN DESIGN

本溪沈溪新城日月岛中央文化区城市设计

用地面积	建筑面积	设计时间	建筑类型
330万平方米	498万平方米	2011年10月	行政中心 商贸中心 文化中心

MEC即复合型国际绿色城市核心区。

传统的城市核心区以高强度的开发塑造现代城市形象，而当下新兴的城市核心区更关注城市环境生态化以及公共空间的合理性，以全新的城市空间布局方式演绎沈溪新城行政中心。文化中心，商务中心的规划设计。"三心并两带"的规划布局把三大中心分解，并围绕城市主干道和主景观道重组，中间围合出多功能共享城市公园。

MEC—multifunctional international ecological urban core area.

Traditional urban core area utilizes intensive development to build the image of modern city, but current emerging urban core area pays more attention to urban ecological environment and rationality of public spaces. This project adopts brand-new urban spatial layout for planning design of administrative center, cultural center and business center of Shenxi New City. The planning layout of "three centers and two belts" divides and reorganizes the three centers around urban trunk road and main landscape to enclose a multifunctional shared urban park.

开发强度

车行流线

公交分析

above   section diagram/局部图解
below   picture of the model/模型照片
right   site plan/总平面图

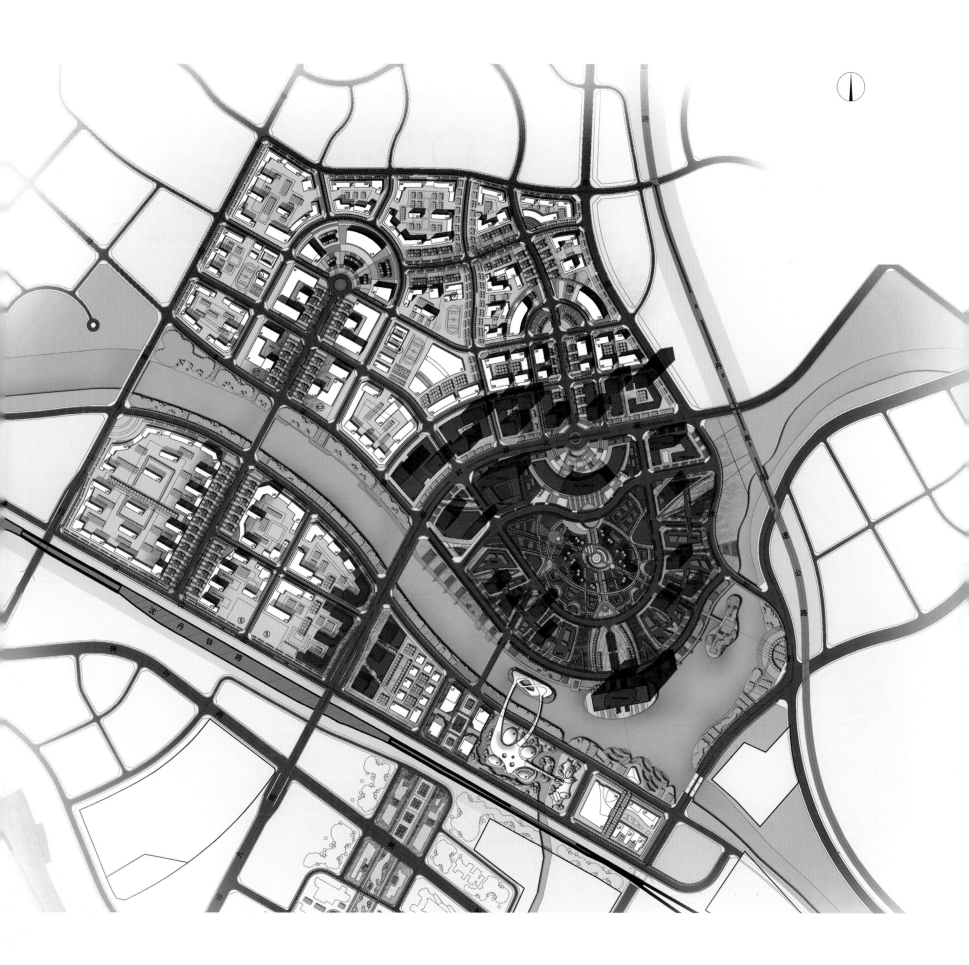

JINAN XIAOQINGRIVER PLANNING

济南小清河规划

用地面积	建筑面积	设计时间	建筑类型
约216.79万平方米	710.81万平方米	2010年	城市规划

本案基地位于济南市中心城区北部，属小清河综合整治片区内，毗邻黄河景观生态带，总体面积约为216.79万平方米，北有二环北路、绕城高速，西靠无影山北路。为城市北跨进程的桥头堡区域，产业以总部经济及现代服务业为主，具有优势型景观资源——黄河。

以特色产业经济活力主轴、区域地标休闲滨河生态空间及生态混和现代社区为规划切入点，建设深绿型低碳城市。引入生态城市主义概念，借由塑造城市的各维度规划要素系统来改变生态流动的方式。

Ji'nan Xiaoqing River Planning Project Planning, located at north of urban center of Jinan City, belongs to Xiaoqing River General Treatment Zone. This project faces to Yellow River landscape ecological belt and covers a total land area of about 2,1679,000m². It is adjacent to Erhuan North Road and Ring Highway at north and Wuyingshan North Road at west. The site is a bridgehead area for northward urban development and mainly consists of headquarters economic industry and modern service industry, possessing advantageous landscape resource—Yellow River.

The planning focuses on featured industrial economy, regional riverside ecological landmark space and modern ecological community complex to create a low-carbon city. The concept of ecological city is introduced and urban planning elements are used to change the ecological flow mode.

三轴二廊五片区：
空间布局上本案以集约式紧凑布局为主要功能排布方式。

南北三轴线：
将沿济泺路、兴隆寺以东道路、河滨规划路形成贯穿基地的三大轴线空间，分别辅以商贸区、文娱区、游憩区的功能配置，将上位规划中散布在基地内部的各项功能用地向此三轴进行集中布置，并成为本案与中心城区联通的重要路径，体现区域新风貌。

东西两走廊：
将二环北路以南第一条规划道路定位为区内主要游憩娱乐轴；将泺口南路以南规划路定位为社区级服务配套轴

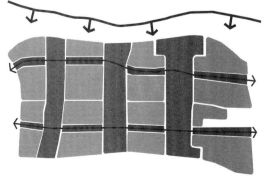

left bird's eye view and structure planning/鸟瞰效果图和规划结构
above site plan/总平面图

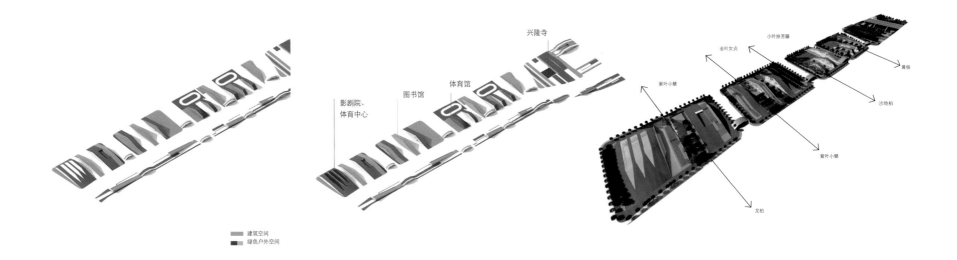

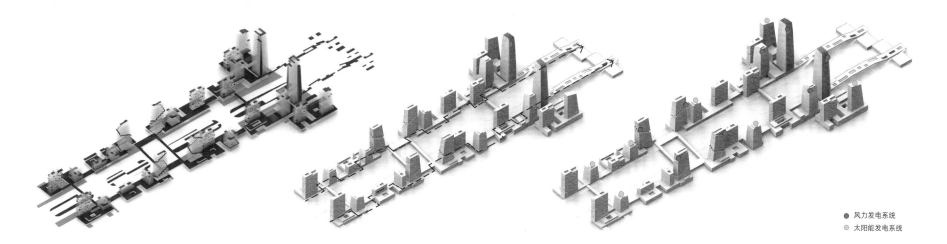

● 风力发电系统
○ 太阳能发电系统

left and right　program distribution diagram/方案功能图解

XI'AN SKY-LAND CITY PLANNIG

西安天地城规划

用地面积	建筑面积	设计时间	建筑类型
297.17万平方米	96.27万平方米	2009年	规划

西安天地城规划区位于古城西安的西北角,与汉长安城遗址的东南边界隔水相望;另一侧靠近二环道路,是连接现代城市发展和历史城市记忆的重要接驳点。

在设计策略方面,我们以"城、路、人、绿、水、墙"六个彼此串联的元素将汉长安城遗址和西安城区连为一体。在重要节点上,大胆规划了以汉韵为根、现代设计手法为茎、中式古典空间关系为枝叶的多个场景,为汉长安城遗址找一个展示的窗口,为未央区造一座活力之城,为西安点一笔添花之彩。

The planning site of Xi'an Sky-Land City is located at northwest corner of ancient city Xi'an and faces to the southeast boundary of Chang'an City relic of Han Dynasty across the river; the other side is near to the second ring road and is an important node connecting modern urban development and historical urban memory.

On the aspect of design strategy, six mutually-linked elements including "city, road, people, vegetation, water and wall" are used to integrate Chang'an City relic of Han Dynasty and urban area of Xi'an. Diversified scenes are planned on important nodes by taking Han Dynasty style as the root, modern design method as the stem and Chinese-style classical spatial relation as branches and leaves. The project will provide an exhibition window for Chang'an City relic of Han Dynasty, a vigorous city for central area, and an outstanding space for Xi'an.

right bird's eye view/鸟瞰图
far right site plan/总平面图

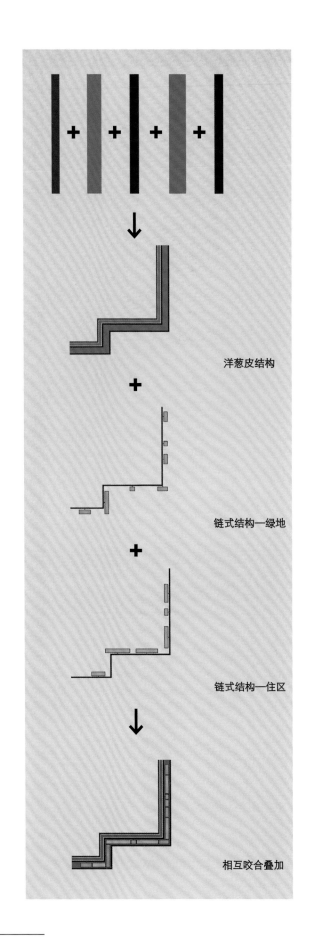
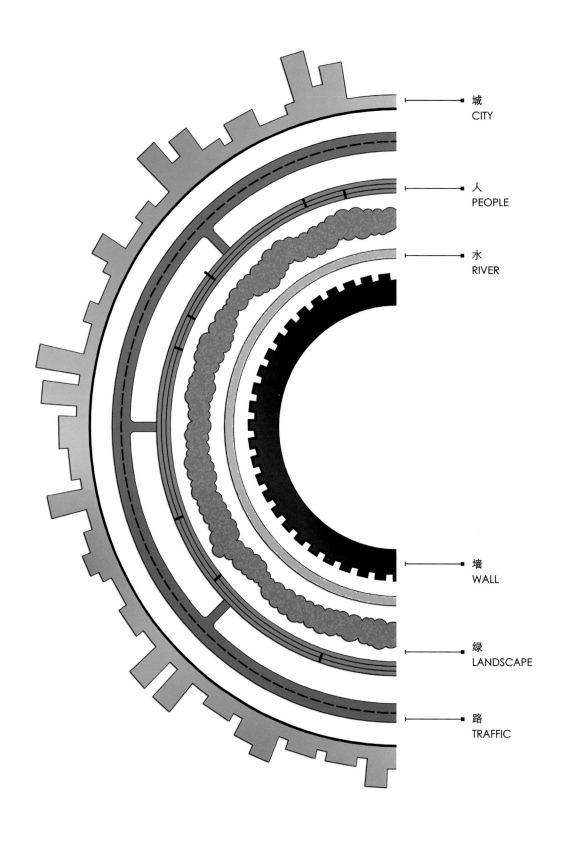

Yun Ying/意向草图——云影

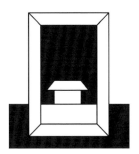

Sky Gate/意向草图——天之门

Cheng Qiang Han Ying/意向草图——城墙汉影

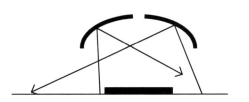

Tian Qiong/意向草图——天穹

left　landscape analysis/景观构筑意向分析
right　Yun Ying Plaza renderings/云影广场效果图

SHENYANG QIPANSHAN PLANNING

沈阳棋盘山规划

用地面积	建筑面积	设计时间	建筑类型
15.94万平方米	15.29万平方米	2010年12月	旅游类规划

沈阳棋盘山国际会议中心项目位于棋盘山风景区内，处于沈阳市东北部，东邻抚顺，北接铁岭，西、南为沈阳市城区，距沈阳市中心约20千米。项目整体布局利用优越的区域环境，将山、水、建筑融为一体，打造"自然的、生态的、健康的、休闲的"国家级会议接待区及山地高尔夫别墅区。贯彻"以人为本，重返自然"的基本思想，创造一个"布局合理、功能齐备、交通便捷、环境怡人、具有文化内涵"的会议接待场所。

强调建筑群体与自然地形的结合，充分利用原始地形的自然形态，灵活布置建筑群体，使其错落有致、空间序列变化丰富。同时丰富了大地轮廓线及区域内部公共空间。使建筑与自然环境有效地融合在一起。把促使人、建筑和环境的和谐共存作为规划设计、建筑设计的根本出发点和最终目标。

Shenyang Qipanshan International Conference Center, located in Qipanshan Scenery Resort at northeast part of Shenyang City, is adjacent to Fushun at east, Tieling at north and urban area at west and south, and is about 20km from urban center of Shenyang. Overall layout of the project utilizes its advantageous regional environment to integrate mountain, water and building together, so as to create a" natural, ecological, healthy and recreational" state conference center and upland golf villa zone. This project adopts the human-oriented and natural concepts to create a conference center with reasonable layout, complete functions, convenient traffic system, beautiful environment and profound culture.

With consideration to natural terrain conditions, buildings are flexibly staggered with each other to realize diversified spatial layout. Meanwhile, ground contour line and internal public spaces are enriched to effectively integrate buildings into natural environment. The fundamental concept and purpose for planning design and architectural design are to realize harmonious coexistance among human being, buildings and environment.

left bird's eye view/整体鸟瞰图
right site plan/总平面图

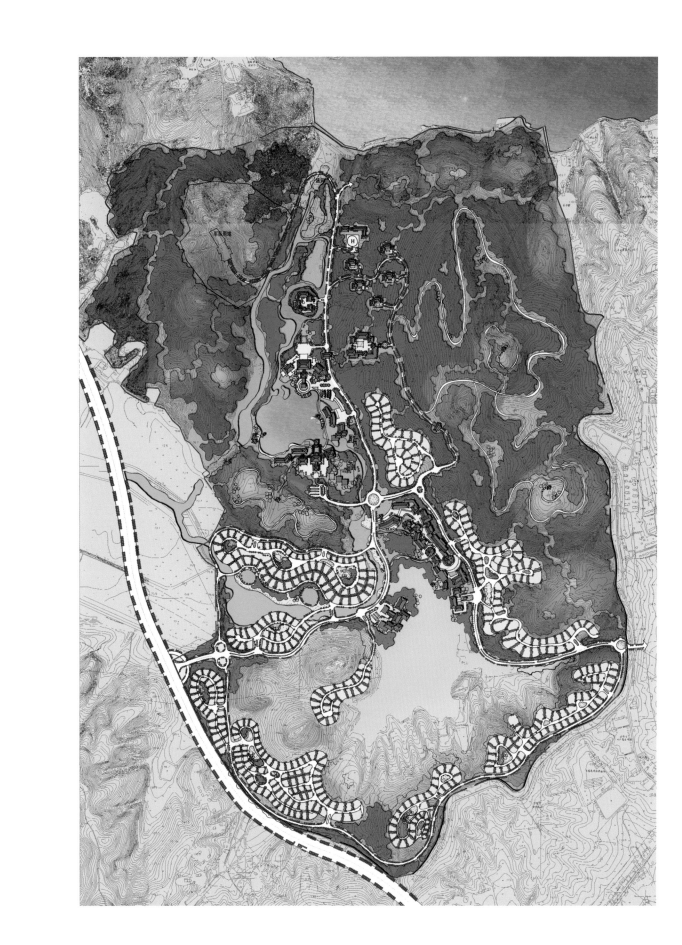

NANJING RIVERSIDE NEW CITY PLANNING

南京滨江新城规划

用地面积	建筑面积	设计时间	建筑类型
1310.19万平米	858.6万平米	2011年	公建

南京滨江新城规划项目是UA与西班牙BLAU事务所进行合作，为业主完成的一项近5000亩大规划项目，UA着重于前期2400亩地块的规划，BLAU事务所则重点研究沿长江风光带及轻轨发展带的设计。

本次规划从对城市尺度、城市扩张模式的研究出发，提出了"指状发展"模式的基本理念；同时与"城市生态主义"相结合，归纳出了"绿络编织"、"交叉渗透"、"端点介入"三大规划理念，将江景通过绿色廊道引入地块腹地，又通过生态公建带与绿色廊道交织，将便捷、生态的生活带到每块用地单元。

Nanjing Riverside New City Planning project is jointly designed by UA and BLAU Firm of Spain and covers an area of about 5000MU (MU is a unit of area, 1MU≈667m²), among which UA is mainly responsible for the initial area of 2400MU, while BLAU Firm pays its main attentions to design of landscape belt and light rail belt along the Yangtze River. This planning is based on research of urban scale and expansion mode to put forward the basic concept of "finger-shape development" mode; meanwhile, it is combined with "urban ecological concept" to generate three main planning concepts, including "green weaving", "cross penetration" and "terminal intervention". River landscape is introduced to the middle part of the site through a green corridor; convenient and ecological lifestyle is realized on each blocks through interweaving between ecological public building belt and the green corridor.

below　site plan/总平面图
top right　connection between the site and Yangtze River/基地与长江的关系
bottom right　bird's eye view from the Yangtze River site/从长江一侧的鸟瞰图

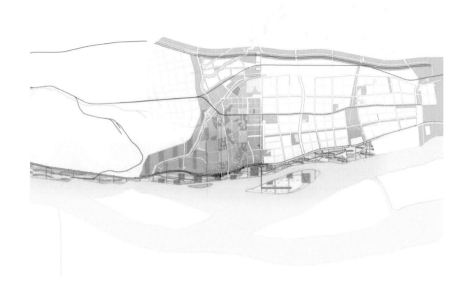

 →	绿络编织 NET	
→	交叉渗透 CROSS	
→	端点介入 ATTRACTION	

EASTERN NEW CITY CBD CENTRAL VIGOROUS DISTRICT PLANNING

徐州东部新城CBD中央活力区规划

用地面积	建筑面积	设计时间	建筑类型
297.17万平方米	96.27万平方米	2009年	规划

徐州城市规划结构为"双核五组团"，其中徐州东部新城位于其中的"一心"，东至京沪高速公路，西抵世纪大道，南临连霍高速公路，北接古黄河风光带，北侧市政府大楼俯瞰整个新城区，是具有包括市级行政、商贸、金融、文化、信息、会展、教育、居住和高新产业的综合性新城。中央活力区位于新城核心地带，总规划面积约为1.13平方千米。规划结构清晰，功能分区合理。结合中央的"都市之门"和"黄金十字轴"布置了中央商务区、休憩商业区、主题公园、开放商住街区，结合围绕在活力区周边的绿色通道和生态水体布置了多个居住组团。

目前多个地块已经建造完成，活力区氛围正逐步形成，在不久的将来必将成为辐射淮海经济区的商务集聚区、服务徐州城市的公共活动中心区和引领徐州新城建设的宜居区。

Xuzhou urban planning layout adopts the concept of "adopts cores and five blocks". Xuzhou Eastern New City is located at one core and is adjacent to Beijing-Shanghai Expressway at east, Century Avenue at west, Lianyungang-Khorgas Expressway at south and ancient Yellow River landscape belt at north. The municipal government building standing at north overlooks the whole new district and is a new city integrating municipal administration, commercial trade, finance, culture, information, exhibition, education, residence and hi-tech industry. The central vigorous district is located at the core area in the new city and covers a total planning area of about 1.13km². The planning realizes clear layout and reasonable functional division. Central business zone, leisure commercial zone, theme park and open commercial blocks are arranged around the central "urban gate" and "golden cross axis"; several residential blocks are designed around green paths and ecological waterscape enclosing the vigorous district.

At present, constructions of several plots have been completed and atmosphere of the vigorous district is forming gradually. This project will soon become a business concentration zone connecting Huaihai economic zone, a public activity center for Xuzhou City and a habitable community leading construction of Xuzhou New City.

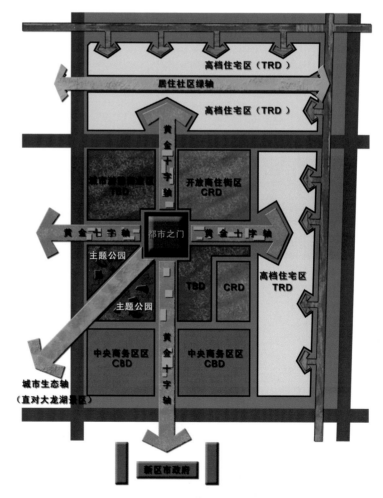

above program distribution diagraml/方案功能图解
right site plan/总平面图

above bird's eye view/整体鸟瞰图
right cultural exhibition area view/文化会展区效果图

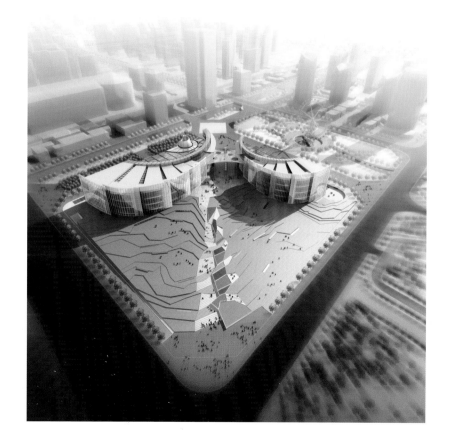

WUHU HIANGHU NEW CITY

芜湖镜湖新城

用地面积	建筑面积	设计时间	建筑类型
95.4万平方米	219.7万平方米	2009年	规划

芜湖镜湖新城位于安徽省芜湖市老城区的东侧,是旧城东扩的触角,是崛起中的城市副中心,担负着承接历史与未来的重任,更是城市更新的一把钥匙。项目通过对城市副中心要素的深入研究,制定了包括建设量、交通、公共空间、配套设施、商办住功能、景观等六大方面的规划导则,将用地分为浴牛塘生态水域片区、国际滨水RBD金融片区、流水社区、开放社区四大片区,其中包括魅力之湖、产业之岛、未来之塔、城市之眼等区域节点。

芜湖镜湖目前已有多个地块建设完成,并将在不久的将来把更多绿色生态和滨水生活方式带给城市。

Wuhu Jinghu New City, located at east of the old urban area of Wuhu City, Anhui Province, is an antenna for eastward expansion of old urban area and an emerging urban subcenter. It takes up both historical and future tasks and is a key for urban renewal. After in-depth research on elements of urban subcenter, planning rules on the following six aspects are estalbished, such as construction workload, traffic system, public space, supporting facilities, commercial, business and residential functions and landscape, etc., to divide the site into Yuniutang ecological waterside zone, international waterside RBD financial zone, flow community and open community, including regional nodes, such as charming lake, industrial island, future tower and urban eyes, etc.

Several plots of Jinghu have been completed and more ecological and watersite lifestyles will be introduced to urban citizen in the near future.

left　site plan/总平面图
below　bird's eye view/整体鸟瞰图

XUZHOU GREENLAND WINDOW

徐州绿地之窗

用地面积	建筑面积	设计时间	建筑类型
36.46万平方米	128.85万平方米	2010年5月	城市规划

本项目位于江苏省徐州市京沪高铁徐州站前。依托基地周边两大生态景区，本规划着力打造城市生态商务核心区，在规划中引入生态神经网络理念，并将这一理念贯穿于城市规划、建筑形态、交通组织、景观设计中，旨在形成独特的徐州城市名片。

Xuzhou Greenland Window Project is located at front of Xuzhou Station of Beijing-Shanghai high-speed railway in Xuzhou City, Jiangsu Province. Taking advantages of two large ecological landscapes around the site, planning of this project aims to create an urban ecological business core area. Ecological neural net concept is introduced in planning and is carried out in urban planning, building shape, traffic organization and landscape design, so as to create a unique business identity for Xuzhou.

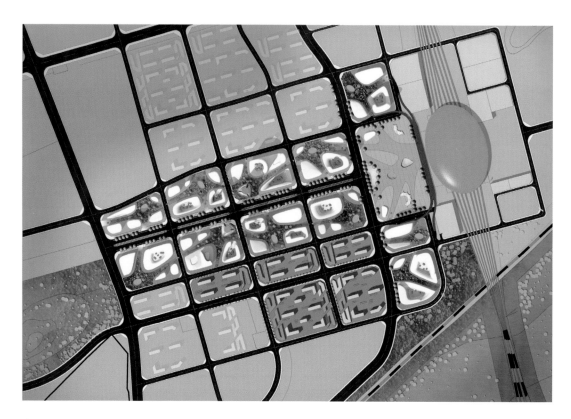

above site plan/总平面图
right bird's eye view/鸟瞰效果图

left   business perspective/商业透视图
top right   bird's eye view/鸟瞰效果图
bottom right   design concept/设计理念

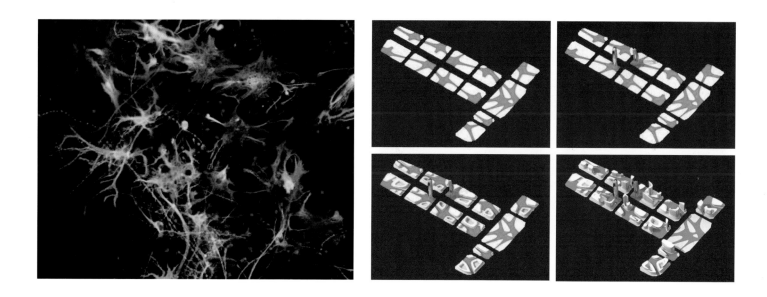